GENERATION AI

Tony Evans

PASSIONPRENEUR®
P U B L I S H I N G

GENERATION AI

Leading People and Machines To Win

Tony Evans

PASSIONPRENEUR® PUBLISHING

Generation AI
Copyright © 2025 Tony Evans
First published in 2025

Print: 978-1-76124-226-7
E-book: 978-1-76124-227-4
Hardback: 978-1-76124-228-1

Publishing information
Publishing and design facilitated by
Passionpreneur Publishing
A division of Passionpreneur Organization Pty Ltd
ABN: 48640637529

Melbourne, VIC | Australia
www.passionpreneurpublishing.com

To my wife, Mariana. Not only for teaching me about Millennials but for the countless hours of walking, talking, listening and debating as I formed my thoughts on this book.

Acknowledgements

There are many people who have influenced my thinking on organisations, on psychology, on AI and on generations.

For Moustafa, my publisher, and his publishing team – thank you for taking me on and for your inspiration. Thank you to Clare and Shobha for all the advice and patience.

The input to the book that I received from people who generously gave time and thought to make it better: Corrie, Kate, Paul, Will, Rasheeqa, Doug, Andy, Ian, Amy. A huge thank you.

My parents, my brothers and my son. Baby Boomers, Gen Xers, Millennials and GenZ. Collectively, from you all I learned so much about the differences and similarities between generations inside and outside of organisations.

For the C-Suite leaders I have spent time with, exchanging ideas on the changing nature of organisations: This has been hugely influential as I assembled this book.

Nick, my GenZ mentor (whom you will read more about in this book): Thank you for generously and patiently giving me your time so I could learn more deeply about GenZ.

Thank you to the team at Aryma Labs, who helped build the GPT that accompanies the book. Working with you is such a great example of how forward thinking companies are using AI to enhance and add to their offering to humanise complex processes.

I am grateful to all the teams I have been part of, the teams of people I have worked with as a consultant, and the teams of people I have managed. I have learned from literally every interaction.

The university that taught me the most about psychology – Heriot-Watt, where I undertook my Business Psychology MSc. Thank you to the faculty for providing such an excellent environment for my initial research into leadership and generations.

The academic work of many people has influenced this book. Particularly influential is that of Jean Twenge, John Kotter, Peter Senge, Geert Hofstede, Erin Meyer, Daniel Kahneman and Amos Tversky.

The biggest thank you to Dr Corrie Block: Without your initial encouragement to write a book, these pages wouldn't be here. And without your influence on my thinking, their content would certainly be of lower quality.

For the many people who made the stories in here possible. Whilst names have in most cases been changed, if you're wondering if the references are to you, they probably are.

And finally, to my primary school teacher Sister Carmel, who said I'd never write a book worth reading. For 40 years you were right.

Table of Contents

Introduction

West of Edinburgh, in the Scottish National Gallery of Modern Art, there is an outdoor installation by the artist Nathan Coley. Some six metres high, lit in neon, it proclaims: 'There Will Be No Miracles Here'.

I first saw this sign after I made a shift in my career to focus on the developing area of digital advertising. As the shift of consumer habits to accessing media online was a seismic one, this new area of advertising was slightly mysterious to most. It seemed to promise a changed future of advertising. Many marketers indeed expected miracles and whilst the performance improvements it brought may have been miraculous over time, the process of getting there was no miracle. It involved a lot of change, false starts, time and resource investment before the shape of the advertising landscape altered and search engines and social media companies became the largest advertising vehicles in the world.

Digital marketing was just one part of the process of digitalisation that has been witnessed over the past few decades. Leaders of organisations navigated this shift, as leaders of organisations have navigated shifts before. The latest shift that the world is seeing is the opportunity of Artificial Intelligence (AI). Whilst this new technology may once more appear mysterious and potentially miraculous, the principle of navigating the team's ideal path will remain the same for the leader as it has been in previous shifts. The extra complexity on this occasion is in the power of the technology and the changes in the people in the workforce.

This book is designed to be a guide to leaders, managers and shapers of organisations as they navigate the shift to Artificial Intelligence. We will discuss current forms of AI and how they can be utilised in the context of your organisation. The main focus of this book is on the people in your organisation, as they are *the* essential factor in successfully integrating AI. Leadership and management approaches have changed over past decades, and must continue evolving in order to get the best out of the people in organisations. We are observing a real difference in approach and attitudes towards the workplace that differs across age groups, which has led to the focus of this book on generations within the workplace.

It is not the first time that differences in the workforce that are generationally based have emerged, but the current gap between the youngest and oldest members of the workforce has never been more stark. Leaders need to work to understand this, as all age groups have an essential role to play in successfully integrating AI. Many current leaders have lived through the transformation of organisations to take advantage of the digital era. By looking at and learning from the mistakes made in the digital transformation era, we can avoid repeating them as we transform our organisations to take advantage of AI.

How the book is laid out

This book is structured to help leaders get prepared for the opportunity that AI brings to the workplace, and to help them prepare for the changes that their teams will need to undergo. This book has been broken down into three parts:

(i) understanding the current state (the essential part of changing anything)
(ii) understanding the people in your teams and how they are changing
(iii) using this knowledge to implement AI for the future.

Part 1 Adapting To The Times	Part 2 The People We Need	Part 3 Stepping Into The Future
The Changing Generations	Group Formation	Hazards of Implementing AI
Modern Leadership	Understanding GenZ	Collaboration and Change Management
The Great Efficiency Illusion	Understanding Baby Boomers	Investing in People
AI in the Workplace	Bridging the Gap	AI Implementation

In the first section, we will look at the current workplace and how it has changed over time. We will look both at how technology and people have evolved over past decades, and how modern leadership has been adapting. We will look at the role of the leader and how this has evolved, and how this has led to changes in how leaders manage others.

We will then focus on what we can learn from the challenges of the last major workplace shift – the shift to digital – and the dysfunctions around data that we see in organisations today. We will view how these have come about and the lessons we can learn from them in order to avoid the levels of waste that were seen in previous digital 'transformations'. Lastly in this section, we will look at the rise of AI, including the opportunity it presents to organisations as well as its limitations and potential problems, along with common misperceptions of what it is capable of.

In the second section, we will look at how groups are formed and the influence of the factors we carry with us that shape our norms and values, such as our age and our national cultural background. We will borrow heavily from both organisational and social psychology to help understand the people within our organisations. Focusing on the two specific groups of Baby Boomers and Zoomers, we will unpack the influences that have led them to their current expectations and understanding of the workplace, and look at how these two groups can complement each other to deliver the successful implementation of AI.

In the next section we will look at the actual implementation of AI – the process of managing change, tactics and strategies for implementing AI, some pitfalls that need to be navigated around and the way to approach developing people in your organisation to prepare for the brave new world. In this section we introduce a framework for the stages of AI integration that companies go through, allowing you to see both where your organisation may currently be and what is involved in the more advanced stages to help you prepare your approach.

Chapter Takeaways and Additional Tools

At the end of each chapter, there are some questions for leaders to ask themselves or their leadership teams. These questions aren't exhaustive, but are intended as food for thought on how ready both you and your organisation are for the changes that are needed.

On my website (www.onv.ai) you will find a link to a Generative Pre-trained Transformer (GPT) that accompanies this book. This GPT has been specifically designed to prioritise the information sources I have used in researching this book as it returns answers. Some of you may already be au fait with using GPTs such as ChatGPT. Along with the 'What Questions a Leader Should Be Asking' section at the end of each chapter, there will be some suggestions of prompts to ask the GPT.

Playing with this GPT gives you an opportunity to ask questions in this format rather than using search. It will also be helpful in understanding the operation and potential limitations of AI. When using it, I would advise looking for two things:

(i) How the quality of the prompt you write gives different outputs
(ii) The limitations of the output.

The first topic is a great one to understand, as this makes all the difference in the output. For instance, try these prompts to see the different outputs:

- 'Tell me about leadership'
- 'Tell me about leadership for someone entering the workplace'
- 'Explain to me as a CEO about the theories of leadership in the past thirty years'
- 'Acting as a PhD researcher writing a paper on leadership theories that have influenced the workplace, explain the transformational leadership theory from Burns and Bass, with additional references for reading provided in a bullet point list. If needed you can ask me questions to clarify the prompt before you answer'

GPT prompting is a relatively new area and we are still working to figure out the ways that get best results. It is not likely that the very first time you insert the prompt it will give you what you want, so be prepared to test by writing, assessing the answer, refining, rewriting etc. I find that including the following in prompts helps:

- Be clear on the goal you are asking for as output
- Tell the GPT to play a particular role
- Provide context to the questions
- Provide detail
- Tell it the length of response you want.

Limitations

If you are not already exposed to GPTs, these models are only as good as the information they are able to access or what they have been trained on. To demonstrate this, this particular GPT has been partially restricted to information relevant to the topics of this book and 'fed' the specific sources referenced in it. Try prompting this particular GPT to 'draw a pie chart of the population of the world broken down by continent', then try the same prompt within another GPT and see the difference in output. At the time of writing, image generation (and in particular text labelling) is a real challenge that AI has yet to overcome.[1]

The restrictions in the data that informs the GPT and the resulting output are a function of the data and the person who programs it, which will lead to both accuracy and bias issues as we will see in subsequent chapters.

Above all else, whether for your own usage or when your teams internally are using a GPT like ChatGPT, take heed of the disclaimer at the bottom of the prompt page: *'ChatGPT can make mistakes. Consider checking important information.'*

---

1   Janelle Shane's blog AI Weirdness has some great examples of animal naming and candy love-heart slogan text challenges that AI has yet to solve: https://www. aiweirdness.com/dall-e3-generates-candy-hearts/.

We live in a time where the rate of change is unlike anything we have witnessed before. The tools available for enhancing our capabilities are advancing at breakneck speed, and at the same time the values and expectations of people in our workforce are also changing. A recurring theme of this book that may give leaders comfort is that the principle of the role of the leader within this era remains the same as with any other navigation of a new opportunity shift that they have gone through before. Assessing the situation, building a vision, developing a strategy, applying resources, measuring impact, managing change, course correcting – the leader is required to employ these skills to successfully lead their team into the AI age.

In this journey, there are many challenges to face and there will undoubtedly be problems to solve. Some leaders may see these issues as something they may be able to avoid by banning AI from operating within their companies. The annals of business history are filled with companies that either didn't recognise the opportunity of new technology or arrogantly believed that they didn't need to change. This book is designed to provide you with the starting points in how to define and adopt the possibilities AI brings within the context of your business, and an understanding of what you need within your teams to make this work. We'll start our journey by looking at the current workplace and how it has changed over time.

PART 1

ADAPTING TO THE TIMES

1

ADAPTING TO THE TIMES

"We cannot direct the wind,
but we can adjust the sails."

– Dolly Parton

"I just don't get why they bloody well insist on inventing these new terms!"

It was a warm summer's day in 2007 and I was sitting with Peter in his office. From the window behind him there was a picture-perfect backdrop of the rolling hills of the green and pleasant countryside of Middle England. Peter was a Marketing Director for an automobile manufacturer that I had known for almost ten years. Now in his mid-fifties, he was struggling to embrace the new world of digital advertising. He was well-versed in the use of media like television, newspapers and posters to position his luxury products and knew that the opportunity of digital media was one he needed to grasp.

We'd just had a meeting in which specialists from the advertising agency in which I worked each presented their activity proposals for the coming year. He asked me to spend a little more time with him after the meeting so he could understand the plans better. Whilst Peter and I were not from the same generation (he was a Baby Boomer, I am a Gen Xer), he sought me out to act as a translator for what the digital specialist (a Millennial) had presented.

Prior to the digital advertising era, Peter was lucky if he got any kind of immediate feedback on how his ads were performing. An ad in a newspaper may generate an action, but it was difficult to tell which of the seven newspapers we had advertised in that week was responsible. All of a sudden his presentation from the digital specialist included 'click-throughs', 'view-throughs', 'dwell time', 'CPAs' (cost per acquired customer) and 'CPCs' (cost per click).

Peter wasn't comfortable raising these questions in the large meeting we'd just had. "I just feel like I'm slowing the meeting down and we've got so much to get through," he explained. "And to be frank, I don't see much downside to just carrying on with the existing ads and not doing more digital. It has worked for us until now!" Peter was certainly sharp-minded – but, like others in his position, very busy. Often, our default nature when we are busy leads us to optimise for something we know already works and that we can make a success of; we tend to deem the alternative risky, even if it offers greater potential. In advertising in the early 2000s, no-one got fired for buying TV slots.[2]

Transitioning to digital advertising was just one part of the digital transformation Peter's company needed to embark on. Digital transformation had huge promise. The collection and successful use of data could inform production, sales and marketing, help make better financial decisions and lead to more delighted customers. But transformation was hard work. Having a vision was one thing – yet galvanising everyone in the company to get behind the idea and work through the required change management is a lengthy process requiring far more than vision. It requires skills in understanding people as well as how to transition and prepare for the cultural change to the company.

This transition to digital in Peter's company was slow and the lack of adoption of digital marketing was just one symptom of it. There are many factors that lead to a company losing market share. Being unable to adapt to changing times is prob-ably the biggest. Peter's firm lost ground to other manufactur-ers who were quicker to adapt and whilst it still manufactures cars, it never regained the market share it lost.

---

2   An adage 'No-one gets fired for buying IBM' was adopted and applied to the adver-tising industry by frustrated digital protagonists when TV ads continued to domi-nate advertising share of spend in the face of other compelling media opportunities.

What Peter didn't see was that back in our office, the issue with understanding the young digital experts was a two-way street. The (Millennial) digital specialist was totally unaware of the fact that their explanations to a valuable client were falling on deaf ears. Whilst Peter struggled to understand our digital specialist, they also struggled to understand the structure, rigidness and decision-making process of the older generation. They had been introduced to computers at school and at home. Their most common way of researching for a college project was more likely to be a search engine than a library. They were avid consumers of digital media – it was their normality. They knew the benefits that it brought in speed and efficiency – and they thought that everyone else knew too.

A Ringside Seat

The advertising industry faced huge upheaval in the first decades of the 21st century as audiences switched from spending their time with 'traditional' media to spending time with digital forms of media, and marketers needed to adapt. A decade of social media growth ensued as Facebook (now Meta) grew from 400 million users to more than 3 billion.

In each evolution, I have been lucky to enjoy a front-row seat as the play unfolded – working in leading companies in digital advertising, in companies using the latest data skills and then for many years at the world's leading social networking company, Meta. In each role, my focus has been to use data to help marketers value what they are buying effectively. You may have heard the adage 'Half my advertising budget is wasted. The trouble is that I don't know which half':[3] Helping marketers to stop wasting money using the application of maths and science was my modus operandi for several decades. Working with many different organisations as a Marketing Scientist, I have had the benefit of seeing first-hand how companies adapted to 'the data age': the structures and processes put in place to try to solve the objective of increasing marketing effectiveness. In so many cases I've come to learn that the use of data is far from perfect. My observation in fact is that almost every company I interacted with is dysfunctional in some way when it comes to data. The majority of this dysfunctionality is due to not focusing enough on the people who need to use the data, relying instead on buying data products or inventing processes hoping that people will adapt to them.

The Injection of Psychology

Among the functions of my job has been to act as translator between generations and manager of people of different ages and backgrounds. The advertising translation of old to new I did for my client Peter (and vice versa to the youngest members of my own team) was something I continued in my role at Meta. Social media advertising was an even greater opportunity to

---

3   This infamous line is attributed to John Wannamaker, a department store owner, nearly 100 years ago. Incredible to think that this wastage carried on making the unscrupulous media operators rich for so long.

reach audiences but required careful understanding of how to interact and interpret the data of advertising performance. I've spent many years managing teams and relationships with different age groups: upwards in age to Baby Boomers and even Silents, and downwards in age to Millennials and GenZers. I've worked to understand the motivations of different people and seen many commonalities between the age groups.

I've always had an interest in group interaction. My deeper fascination with personal interaction, social interaction and team dynamics was first ignited by psychological experiments conducted by people like Stanley Milgram (1963), Philip Zimbardo (1971) and Solomon Asch (1951) that I'd read about when I was in my teens.

The results of these research studies into obedience and compliance to a group norm most likely lit the fire of my interest in advertising for the first part of my career. In my roles managing Gen Xers, Millennials and GenZers, I learned about how the age groups differed in approaches to work, the tools that were used in the course of executing work, and the different expectations people had in team interaction. To understand as much as I can of the youngest generation, I have a mentor, Nick, whom I will refer to within this book. Nick is 21. I would highly advocate this form of mentoring if you truly want to understand GenZ. All too often we look to mentoring by selecting someone 'senior' and either peer age or older. Nick is certainly senior to me in knowing how his generation sees the world, what they want from the workplace, and vitally – why.

My own yearning for a better understanding of human behaviour in groups and group dynamics in these situations took me back to school to fill the lockdown evenings during the pandemic, researching psychology as it relates to business. The theories of psychology I studied gave me invaluable insights: different social groups and how they operate, change management, types of leadership, responses to leadership, methods of interaction, the psychology of coaching. All of these are things I wish I'd learned 20 years early in my career. I believe that every new and rising manager as well as leaders would benefit hugely from understanding these theories.

Finding Our Place in the World

A simplified way of looking at ourselves is as a product of our environment and the processing of our experiences within that environment. Psychologists have long noted the influence of the early years of our lives. We have many influences as we make the transition from being purely self-centred as infants into learning how to interact successfully (or not!) with our environment and others within it. Gaps in understanding others are often based on different frames of reference which are a product of our experiences in life.

Working in global organisations and managing teams of different ages and from different countries, I've observed that two of the largest influences on the way we interact with our work environment are the era in which we grew up and the national culture that we grew up in. From the era in which we grew up, if you grew up in a time where food was rationed and the quickest way of sending a message to someone was via the post, you have a different lens to see the world through compared to someone born in abundant times who has always been able to send and receive messages *to any part of the world* in mere seconds.

The influence of national culture has fascinated me as I've travelled and worked with people in a number of different countries. As someone who grew up in England, I have inherited a range of deep-rooted beliefs about how I should behave in society. The thought of jumping a queue for the bus for instance fills me with deep dread. Similarly, the English also have a tendency to be indirect in resolving a situation when someone flouts the unwritten rule of 'no queue-jumping'. To address the perpetrator directly seems to evade the English, and most would prefer to either comment to the person next to them or if really frustrated exclaim rather loudly "There is a queue you know!" rather than seek the person out and explain their displeasure. Similarly, not to be acknowledged when holding the door open for a stranger as an Englishman will generally not result in a confrontation, but a loud and sarcastic "Thank you?!" will be uttered after the non-thanker has passed.

Every society has these subtle expectations that we absorb by being a member of that culture. The acceptable level of volume of a public conversation. The level of personal space we expect. One, two, three (or even more) kisses in a greeting. Whether a gentleman should ever wear speedos in a swimming pool or on the beach. All of these are subtly implanted into us by our immediate social groups in our impressionable developmental years.

One of the most beneficial learnings I've had when managing teams from different national cultures has been about the impact of these cultures on our expectations in the workplace. In my own studies and practices I've drawn heavily from the work of Geert Hofstede (2001), a social psychologist.

Hofstede's work maps countries on a number of dimensions: (i) relationship with authority, (ii) how individual or collective the society is, (iii) the motivation for achievement and success, (iv) the appetite for risk/risk avoidance, (v) how short term or long term they typically think, and (vi) how 'indulgent' they are (the attitudes towards enjoying life and having fun). It is the comparison between countries on these dimensions that can show up stark differences in the values that a person from a company might have. A wonderful tool on their website allows for a look across these dimensions in countries around the world, enabling us to see how an average person from each country compares to people from other countries.[4]

Presuming that everyone from one country or generation is exactly the same is of course a trap to be avoided. The social

---

4   This excellent tool, which anyone can play with, can be found at https://www.
     theculturefactor.com/country-comparison-tool

norms in a national culture are however large influences on a person's life, and understanding the impact this has is valuable. From a personal point of view, working with teams to explore the impact of national culture, I've witnessed real 'penny drop' moments when sharing the common traits displayed by different countries. On revealing these, I've seen team members suddenly see each other in a new light. The discovery for instance that it is merely the influence of certain national cultures that creates a person's tendency to be direct was a revelation, as some had previously considered it rude. Or that national culture may be informing an individual's higher appetite for taking risks, which previously had quietly been considered reckless. These things can help break down some invisible barriers in a team.

The Rise of AI

The big moments that have played out in my career are among the pivotal ones that have shaped the world of industry that we know today, and these moments impacted many people in the workforce as have previous revolutions. The Industrial Revolution, technological revolution, digital revolution and now the advent of Artificial Intelligence: All of these brought challenges to those leading companies through the changes.

Operating for more than a decade within a company that was an early adopter of and a heavy investor in AI (Facebook, now Meta), I gained a good perspective on what AI is and what it isn't – and whilst it might be powerful, it is not magic. It has already been used to great impact, and we've witnessed PR disasters from not having proper guardrails in place in how to integrate and adopt it.

The complexity and capability of the human brain far, far eclipses AI and will do for many years to come. But AI can do great things in creating efficiencies of certain tasks, is already impacting our lives and is of great benefit to organisations that can harness it. AI implementation and successful adoption however needs people – and people are way more difficult to make work the way you'd predict than machines! Leaders have a particular challenge to contend with in the world of AI. The fundamentals of leading a business through this change are the same as other business transitions: decisions on how to best use resources to align a group of people behind a collective goal; how to evaluate progress, course correct, reiterate. The people that are needed within our organisations are where there are the largest gaps. The differences in values, expectations and motivational needs has never been more stark across the ages of people we employ.

The leader's role is to pull this all together into a beautiful harmonious and high-performing unit. This means that leaders need to understand this new technology's potential along with the skills needed for successful adoption, then bring together the different skills that are possessed by different groups of employees. In the next chapter we will look more closely at the different generational groups in our workforces.

GPT Prompts

- Provide a clear breakdown of the current workforce composition in [COUNTRY OF INTEREST], categorized by generational groups: Generation Z, Millennials, Generation X, and Baby Boomers. Then, offer well-sourced projections for how this generational mix is expected to change by 2030. Present the information in a format suitable for executive briefings—ideally with a chart or summary table.

- Act as a strategy consultant preparing a board-level report for a [AREA OF COMMERCE] company. Summarise the following: 1. Key shifts in consumer behavior over the past 2 years 2. Emerging supply chain risks or constraints 3. Marketing and digital channel trends relevant to the industry Also, outline recent or upcoming regulatory changes affecting operations in major global regions (e.g., North America, Europe, Asia-Pacific). Structure the output in executive-summary style, with bullet points and headings.

What Questions Should a Leader Be Asking?

- How could AI change the employer–employee relationship?

- What trends and disruptions have emerged in our industry that have impacted us over the past three years?

- How can I ensure that my approach to leadership remains relevant in managing both technological advancements and the evolving needs of a multi-generational workforce?

- What steps can I take to help my team navigate the potential challenges and fears associated with AI, ensuring smooth adoption across all groups within my organisation?

2

GENERATIONAL DIFFERENCES

Fearing For Your Children's Children

In this chapter, we will look more deeply at how the era in which we were born influences our team's approach to the workplace, and how that in turn influences their relationship with AI.

The interest in attitude difference by age started for me with music. As a child growing up in Britain in the 1980s, I was spoiled with exposure to a wide variety of different music. Each decade's popular music was connected to social topics of the era, and the 80s had a defining musical sound that was different to the decades before and those that followed.

I became very aware of these social groups, not just in music but also in other collective interests, whether music, fashion or a particular sport – even fighting those from other groups on a Saturday afternoon. Some of these groups were organised and had fixed rules or expectations for participation, while others were casual in nature.

Alongside these groups were other groups that I belonged to as I went through my teens: family groups, school, religion, being a member of a village or town, being part of a country group, becoming a taxpayer, and starting to vote. Being part of these groups forms part of the identity we carry with us through our lives.

Some groups seem to form effortlessly, while others are more complex. Our workplaces are particularly complex social structures, as their complexities stem from we humans who inhabit them. Each person has their own understanding of the world, formed by processing a set of experiences unique to them. These differing experiences lead to us forming values and norms. If not discussed or not resolved, these differences can often lead to great conflict when brought into the workplace.

The Occupy Generation

I was aware of the different attitudes often held by people of different ages, but it wasn't until I met Jon, an Occupy protestor who thrust a leaflet into my hand as I walked past London's St Paul's Cathedral in 2011, that the generational divide *really* started to sink in. He was one of many young people involved in the protest, where tents were camped outside landmark buildings. We came from remarkably similar backgrounds. He was perhaps 15 years my junior, educated in a similar way to me, but his prospects for the future looked much dimmer as he had suffered much more than I had as a result of the 2007 economic crisis.

This protest wasn't restricted to a few 'raving reds', as my Milton Friedman–loving[5] economics teacher would have called them. With the benefit of tools like smartphones and social media, both ideas and the ability to coordinate with ease helped the protests spread to over 82 countries and over 950 cities globally (Wikipedia 2024, Taylor 2011). There was a very real difference between how my life had unfolded, and how Jon's and countless other people's lives had unfolded, purely due to the year in which we happened to be born. It was this conversation that stuck with me, helping me begin to understand the era in which we were born as a major influencer of our behaviour.

The Worry About the Next Generation

About the same time, I was at a retirement party for a friend of my parents who was leaving his role as a school headmaster after more than four decades educating children.

When I asked him what he thought about the children currently in his pastoral care versus how children were when he

---

5   A notable economist with some theories that were depressing in the 1970s and damaging for society for the next 40 years.

first started as a classroom teacher, he quoted to me in his booming, headmastery voice the Latin phrase, 'Si ita est posteritas timeo nepotum'.

He explained that this was an example of early graffiti from ancient Rome, found scrawled on a wall in an archaeological dig. The English translation is roughly, 'If this is the next generation, I fear for my children's children'. The role of elder generations throughout history always seems to be thinking that the next generation is worse.

Time has shown us that the next generation are no worse, just different. For those who have been in the workplace the longest, things are certainly very different to when they first entered. The distance to the newest entrants to the workplace certainly feels more significant to the longest-serving than anything they may have witnessed in their careers. The technology used, methods of communication, attitudes towards the workplace and speed at which complex tasks can be accomplished looks very different to 50 years ago.

The Impact of Birth Era

'Generations' are defined as 15–20 year periods into which people were born. Whilst disagreements exist on whether whole groups of people can be defined simply by the years they were born, most accept that growing up in the post–World War II era of the 1950s and 1960s leads to a very different set of experiences than growing up in the early 21st century, resulting in different expectations and values.

The commonly accepted definition of generations relevant to the people we meet in the workplace is:

- The Silent Generation (1925–1945)
- Baby Boomers (1946–1964)
- Generation X (1965–1980)
- Millennials (1981–1996)
- Generation Z/Zoomers (1997–2012)

The proportion of each group in the workforce varies by industry but broadly within current workforces, 20–25% are GenZ, 35–40% are Millennials, 30–35% are Gen X, and 10–15% are Baby Boomers. These numbers are rapidly changing: By 2030, GenZ is forecast to be the largest generational group in the workforce.

Exposure to available technology is a major factor that influences development within each generation. For instance, the access to machinery to complete domestic chores more simply, such as cooking on a stove versus an open fire or washing clothes in a machine, shaped the choices of work that Baby Boomers were able to undertake.

The ability to travel long distances by air; the access to birth control, enabling parents to choose when to have children; and the ability to remain connected in real time to friends via social media are all examples of technology that have had seismic impacts on how generations have developed. What age we are when a major global event occurs also defines its impact on people of different ages. For instance, the economic crisis of 2007–08 and the COVID pandemic in 2020–21 had a very different impact on people who were entering the workplace versus those who were about to retire.

Boomers and Zoomers

The two groups we will focus on the most in this book are those most established in the workplace (Baby Boomers) and the newest entrants (Generation Z or Zoomers[6]). It is these groups that have the skills needed to successfully implement AI within the workplace: Zoomers are the group adopting AI the most within their day-to-day lives, while the Baby Boomers have the experience of witnessing new technology implementations and PR disasters. Their understanding of ethical considerations and ability to measure risk comes from experience. For the most successful integration of AI in your business, these two groups need to be aligned – yet they currently sit worlds apart.

---

6   'Zoomers' was the moniker bestowed upon the younger generation who grew up using the popular video conferencing technology Zoom during the Covid-19 pandemic.

The Speed of Progression

Technological development has had a huge impact on how we live today compared to even 20 years ago. Moore's law was first defined in 1965 and has stood the test of time very well, correctly foreseeing how technology would develop. It states that the capacity and speed of computers can be expected to double every two years due to the number of transistors an integrated circuit can contain. After the laws of physics get in the way and the physical size of transistors cannot be reduced further, a different type of technology will undoubtedly continue this path of progression. Whether quantum computing, photonic computing or carbon nanotube transistors,[7] progress is inevitable. To many of us, this progress is what is marvellous about being human.

Yet at the same time as this rapid technological change is occurring, some things are *not* changing quickly. Our brain's methods of processing information, the generation of emotions, how interactive relationships form and the way we learn about our participation in groups is a process of evolution over millions of years. The last 50 years of technological development, or the last few centuries of organising our collective skills into companies to provide a return to that collective, is a pretty small dot on the timeline of human development.

---

7   Professor Leiserson and Dr Schardl of the Supertech Research Group have strong opinions here; see Woods (n.d.).

GENERATION AI

The forming of social groups to achieve a collective aim is a key part of participation in modern society. Within most organisations, there is a diversity of backgrounds, ages and approaches to how work gets done. Being able to successfully bring AI into your business requires reliance on the whole of the organisational group.

Leadership in the AI Age

The potential of Artificial Intelligence has captured the imagination of writers and film directors for many years. The potential for a non-human intelligence to wreak havoc on what it means to be human plays on our fears, but the potential for such intelligence to enhance our lives is enthralling.

The development of AI is not new, having its roots in the middle of last century. Alongside AI being used for solving mathematical problems, an early chatbot was developed in the 1960s. Computers have also been beating humans at games for some time.[8] We have had artificial robot assistants for domestic chores since the Roomba entered homes in 2002, while the business world began adopting them in 2006, when companies like Facebook (now Meta), Twitter (now X) and Netflix (still Netflix) started using AI.

---

8   In 1997, Grand Master Gary Kasparov was beaten by IBM's Deep Blue, the first computer to beat a world chess champion.

The Leader's Role In the AI Era

Also not new is the principle role required to be performed by business leaders, which has changed very little since we started forming companies. Leaders harness available resources, applying these resources to a task for the collective good of a group. They then assess and refine the application of the resources, make adjustments and repeat.

The methods of management have changed somewhat over time. Thankfully, the way we think of the role of business leaders has evolved. With the benefit of research and greater access to information, we see that successful leaders today have become well-versed in understanding the needs of individuals in the group.

Understanding these needs is a psychological process, because people are not 'resources' that can be applied in the same way as raw materials are. Instead, each of us has a set of needs, wants, hopes, dreams and desires that accompanies us on our journey through life and operates as a driver of our behaviour. We also carry with us our own norms and values, influenced by our experiences, many of which happen in our formative years of life via our interactions with other members of our social groups. It is the interaction between people within our businesses that is at the core of our organisations being success-ful. The importance of the role of people in organisations will continue – we will likely see roles changing or being enhanced by AI, but the chance of eliminating humans from the process of organisations is very slim indeed.

Successful AI Integration Must Unite Generations

For leaders starting out trying to navigate the hot topic of AI, one factor to be understood is what it is capable of and what it is not. How to manage the change that it will bring within your company and how to motivate your people to make the most of it will depend on your ability to understand the complex and differing needs of subgroups within your organisation and how to bring these groups together to utilise their complementary skills.

The dynamic relevant to leaders right now is that the workplace is made up of some people with very different needs and worldviews. There is much research on and many books written about human psychology. Two of the most fundamental elements we will use in this book as major influences on our values are the era in which we are born, and the environment in which we spend our early years. In the multi-age and globalised workforces that exist in many workplaces, it is the understanding of these dimensions that will give leaders a huge advantage, whether in the shift to AI or in the many other elements that impact their business.

Summary

The role of leadership is the same as it was last year, and the same as it was 50 years ago. Artificial Intelligence is just the latest opportunity for improvement in our organisations. Implementation of AI, of course, involves investing in new technology, but the more important consideration is about managing the change it will bring. Although this is not as obvious as purely training people to use a new technology, understanding the subgroups who will use it and how to bring them together is crucial.

Leaders and aspiring leaders aiming to navigate the successful adoption of AI must focus on people in their organisations, each with a unique blend of skills, approaches, levels of knowledge and experience. They also have different life experiences, values and norms that shape their approach to work. Perhaps the most contrasting of these can be found at the extremes of age groups – the youngest entrants to the workplace and those who have been at it the longest. How to hire, motivate, grow and help through periods of change is the challenge for the leader in the AI generation. It is this focus on *people* – not products or processes – that will enable successful adoption. Artificial Intelligence is the new vehicle that leaders need to embrace, but with AI being attached to so many company launches, product offerings, company statements and sales pitches, it's easy for anyone to get confused about what it actually is. Let's start by defining AI ...

GPT Prompts

- Provide an overview of the major technological influences on generational groups over the last 50 years in both home and workplace scenarios, including references to academic experts for deeper learning of societal impact.

- Compare how the COVID-19 lockdowns affected two distinct employee groups: (1) individuals in the first year of their professional careers, and (2) individuals nearing retirement. Focus on workplace experience, mental health, digital adaptation, and career outlook. Support your summary with reputable sources or studies CEOs can reference for further insight.

What Questions Should a Leader Be Asking?

- How have the different generational experiences in my team shaped attitudes to work and new technology in my organisation?

- How well do I understand the unique values, motivations and norms of the diverse age groups within my workforce?

- How is my organisation currently fostering collaboration between older and younger generations to leverage their experiences and expertise?

3

AI IN THE WORKPLACE

Is There More to AI than ChatGPT?

"Oh, is this the way they say the future's meant to feel?
Or just twenty thousand people standing in a field?"

— JARVIS COCKER/PULP

Aclear understanding of the role AI can play must be established by leaders before they can build their vision and mobilise their teams around it. This chapter is intended to help leaders as they think through the integration of AI into their organisation's working practices. We will be looking at the breadth of AI and exploring its potential as well as its limitations. Aligning your organisation on AI's true nature is a necessary cornerstone on which everything from strategy to investment to operational practices will be built.

Enter ChatGPT

It was a late evening in November 2022 when Mansour first consciously interacted with AI. As he passed a young colleague's desk on the way out of the door, he saw his colleague's face literally agog as he stared at his screen.

'Have you seen this??!' his colleague exclaimed, eyes glued to his screen. Mansour could sense something big was occurring from the tone – perhaps a restructuring announcement or the resignation of a board member or news within the financial markets. But as he looked over his colleague's shoulder, it appeared to be a chatbot.

'I just asked it to rewrite my presentation notes in the style of Snoop Dogg. And it got the tone bang on. So I asked it to rewrite it in the style of Aristotle. And it did that too. I've just asked it to come up with a speech for my brother's wedding. All of it done in seconds!'

Mansour recounted the story to me of how he first discovered it, and how by the end of the next day as they had shared with other colleagues and played with it more, ChatGPT was condensing emails into bullet points for ease of reading, summarising outputs from an Excel spreadsheet, helping source information for some MBA coursework, writing a training plan for onboarding new recruits, assisting with some school physics homework and even writing a Python script for one of the data engineers.

In almost all of the cases the work needed refining, but the output amazed everyone who interacted with it. It seemed that their imagination, plus the ability to ask the question in the right way, was the only thing that limited the output. Whilst it looked and acted like a chatbot, ChatGPT offered something much more powerful than anything the general public had previously had access to. The chat interface was based on large language models (LLMs), giving the public their very first taste of interacting with a Generative Pre-trained Transformer (GPT).

AI For the Masses

Five days after the release of the free version of ChatGPT, it had been downloaded one million times. By the end of January 2023, it was estimated to have 100 million active users (Hu 2023).

For many people, this was the first time they had consciously engaged with Artificial Intelligence. This moment heralded a boom in the conversations we were having in our homes and offices about a tangible application of AI. Up until this point, to many people, AI had been the topic of science fiction films which painted a grim picture of a dystopian future where the machines had become sentient and were busily exterminating humans as a threat to their existence.

Whilst this point marked a milestone in the power of making Generative AI accessible to a wide audience, AI is not new. The

Turing Test,[9] which aimed to demonstrate whether machine intelligence was present, was first proposed in 1950. The term 'Artificial Intelligence' was first coined by computer scientist John McCarthy in 1955 and over the following decades, in amongst the false starts and dead ends, milestones were hit where computers exceeded human performance in games and indeed beat the famous Turing Test. Whilst the world of computer science and those working in the technology field have been aware of the use of AI for some time, even the average person has quietly seen AI influencing their lives in some way for at least the last decade. With everything from recommendations of which film to watch next on Netflix and buy next on Amazon to ads on social media and coupons emailed from the local supermarket becoming more relevant, AI has become more sophisticated and personalised as companies increasingly embrace it.[10]

The following diagram depicts the key moments in computing history of AI's evolution.

---

9   The Turing Test was designed to test for intelligence in a computer. In order to pass the test, the answers between the computer and another human needed to be indistinguishable.

10 If you need further convincing of the power of AI-driven personalised recommendations, try logging into your wife/husband/child's Netflix account and seeing the difference in what films and programs are recommended.

FIGURE 1 – ARTIFICIAL INTELLIGENCE TIMELINE

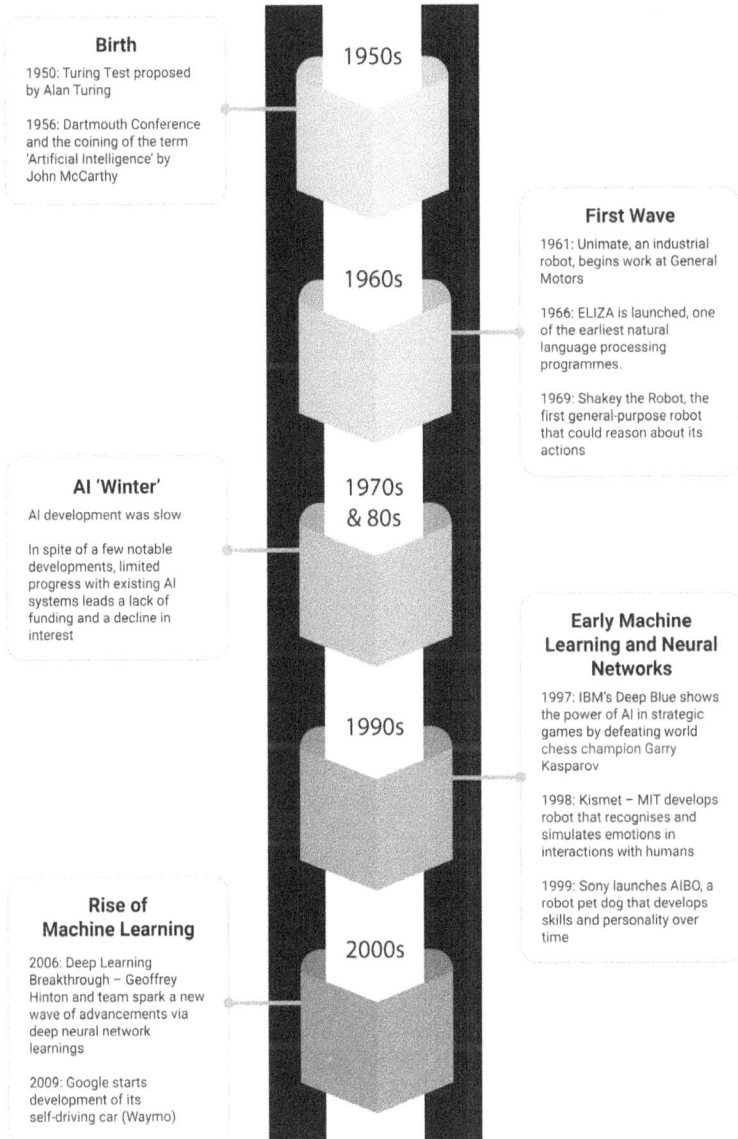

1950s

Birth

1950: Turing Test proposed by Alan Turing

1956: Dartmouth Conference and the coining of the term 'Artificial Intelligence' by John McCarthy

1960s

First Wave

1961: Unimate, an industrial robot, begins work at General Motors

1966: ELIZA is launched, one of the earliest natural language processing programmes.

1969: Shakey the Robot, the first general-purpose robot that could reason about its actions

1970s & 80s

AI 'Winter'

AI development was slow

In spite of a few notable developments, limited progress with existing AI systems leads a lack of funding and a decline in interest

Early Machine Learning and Neural Networks

1997: IBM's Deep Blue shows the power of AI in strategic games by defeating world chess champion Garry Kasparov

1998: Kismet – MIT develops robot that recognises and simulates emotions in interactions with humans

1999: Sony launches AIBO, a robot pet dog that develops skills and personality over time

1990s

2000s

Rise of Machine Learning

2006: Deep Learning Breakthrough – Geoffrey Hinton and team spark a new wave of advancements via deep neural network learnings

2009: Google starts development of its self-driving car (Waymo)

45

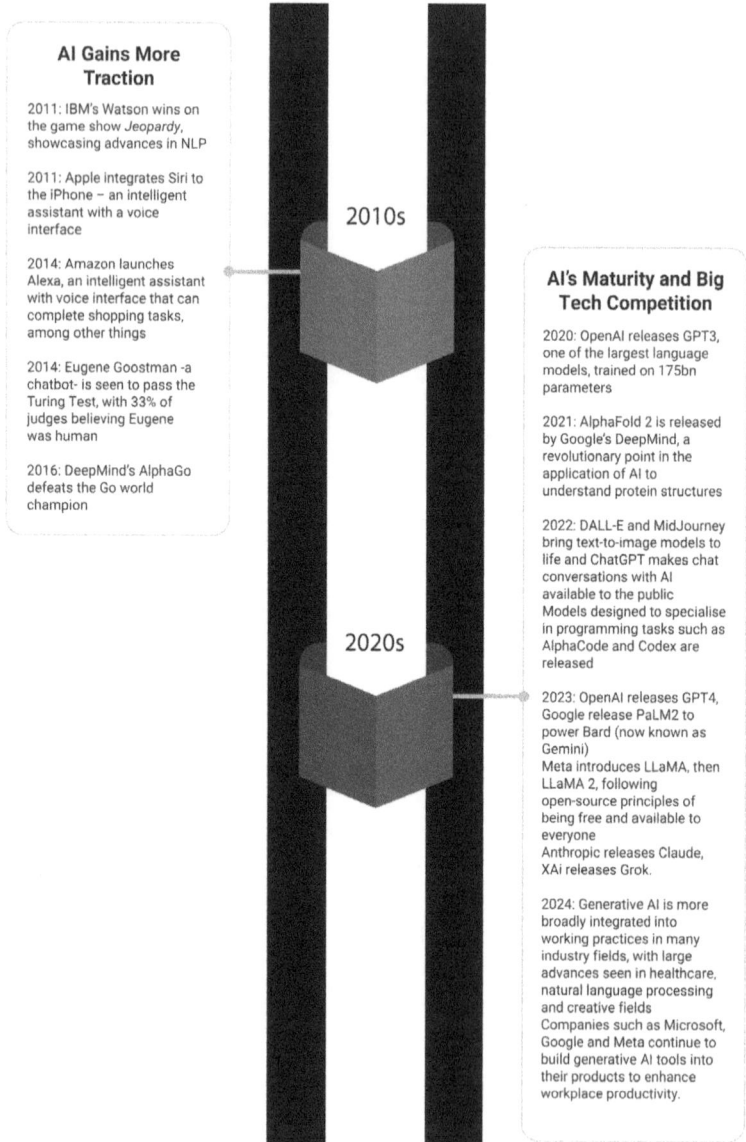

AI Gains More Traction

2011: IBM's Watson wins on the game show *Jeopardy*, showcasing advances in NLP

2011: Apple integrates Siri to the iPhone – an intelligent assistant with a voice interface

2014: Amazon launches Alexa, an intelligent assistant with voice interface that can complete shopping tasks, among other things

2014: Eugene Goostman -a chatbot- is seen to pass the Turing Test, with 33% of judges believing Eugene was human

2016: DeepMind's AlphaGo defeats the Go world champion

2010s

AI's Maturity and Big Tech Competition

2020: OpenAI releases GPT3, one of the largest language models, trained on 175bn parameters

2021: AlphaFold 2 is released by Google's DeepMind, a revolutionary point in the application of AI to understand protein structures

2022: DALL-E and MidJourney bring text-to-image models to life and ChatGPT makes chat conversations with AI available to the public
Models designed to specialise in programming tasks such as AlphaCode and Codex are released

2020s

2023: OpenAI releases GPT4, Google release PaLM2 to power Bard (now known as Gemini)
Meta introduces LLaMA, then LLaMA 2, following open-source principles of being free and available to everyone
Anthropic releases Claude, XAi releases Grok.

2024: Generative AI is more broadly integrated into working practices in many industry fields, with large advances seen in healthcare, natural language processing and creative fields
Companies such as Microsoft, Google and Meta continue to build generative AI tools into their products to enhance workplace productivity.

The Task for Leaders Right Now

Building a company that can bring the benefits of AI into enhancing daily operations is a significant undertaking. It requires a bold vision, strong leadership, carefully crafted strategies, a regulatory framework conducive to digital transformation and a technology infrastructure that enables this objective. It's also crucial to have a collaborative ecosystem of government and private sector entities that build talent, deploy AI technologies and drive adoption.

Reassuringly for the leader, in amongst the debate of whether AI will add jobs or remove them, AI is *not* going to take the job of business leadership! As with any other business opportunity, leaders will need to understand how to put resources into AI integration, align teams around it, measure the impact of the resources deployed, course correct, refine and reiterate.

Before any resources can be deployed though, understanding and defining what AI actually means to the organisation is a necessary first step for leaders in mobilising the appropriate people and teams. On one side, there are claims that AI is about to revolutionise the way we work and is destined to make many jobs defunct within a couple of years. On the other side, there are claims that AI is an unsophisticated tool of limited practical use to organisations. Somewhere between these two statements is the truth for most of us: Some industries are going to benefit from AI sooner than others. The timescale of how soon this disruption will happen to each industry is the debatable point. It is unlikely to change everything you know any time soon, but companies that fail to deploy and adapt to AI within this decade will likely be the most disrupted.

How Should Organisations Define AI?

Most leaders are still at an early stage of understanding the potential impact that AI will have on their workplace. For many people, interactions with AI may be limited to generative AI programs like ChatGPT – but AI covers far more.

Understanding what AI is and isn't, and then determining how the many different forms of AI can be used to transform business practices, is the practical first step for leaders.

Simply defined, AI is a technology that enables computers to simulate human intelligence and develop problem-solving capabilities. The term can cover a broad range of meanings, depending on which member of your team you are talking to or which company is trying to sell you something. It encompasses many different fields and applications of computer science. For example, some common terms currently in use (certainly not an exhaustive list) include machine learning, deep learning, generative AI, natural language processing (NLP), large language models (LLMs), foundation models, computer vision and automation. Each of these categories can then have many sub-terms to classify activities further. Within machine learning, for instance, there are terms for algorithms and other applications such as Random Forest, K-Means Clustering, Neural Networks and Gradient Boosting Machines. The terms you hear may even vary by business sector. A list of common applications of AI currently in the workplace can be found in Table 1.

TABLE 1: TYPES OF AI AROUND US DAILY

Chatbots and Virtual Assistants	AI systems that can simulate human conversation, providing customer support and automating routine tasks.
Predictive Analytics	Utilised for forecasting future events or behaviours by analysing historical data, informing models for applications such as dynamic pricing.
Image Recognition	Algorithms that identify and classify objects in images, with use cases existing in many fields such as security, healthcare diagnostics and retail.
Natural Language Processing (NLP)	The processing and understanding of human language, enabling applications like sentiment analysis, language translation, and content generation.
Recommendation Systems	Analysis of user behaviour to suggest products or content, significantly enhancing user experience in social media, e-commerce, and streaming services.
Autonomous Vehicles	AI systems that enable self-driving cars, using sensor data for navigation and decision-making.

Fraud Detection	AI algorithms that detect unusual patterns and potentially fraudulent activities, particularly useful in finance and cybersecurity.
Robotics	AI-driven robots perform tasks in manufacturing, logistics, and even surgical procedures, improving efficiency and precision.
Personalised Marketing	AI that tailors marketing content to individual preferences, optimising engagement and conversion rates.
Speech Recognition	AI that transforms spoken language into text, enabling voice-activated control systems and transcription services.
Generative AI	AI that is capable of generating images, videos, text or other forms of data using generative models. By learning patterns and structure from training data, it produces new data that is similar.

Is Defining AI That Difficult?

For the curious, the explanation of how AI actually works in any of these areas is in itself a challenge. Technology historian George Dyson has written extensively about the developments of Artificial Intelligence. In his essay of the same name, he posits the term 'The Third Law' about the complexity of Artificial Intelligence. The Third Law states that 'any system simple enough to be understandable will not be complicated enough to behave intelligently, while any system complicated enough to behave intelligently will be too complicated to understand' (2019).

There are certainly many complexities that exist within the categories and subcategories that attempt to define AI. Amongst those shaping this field, whether innovators, regulators, investors, business leaders or philosophers, the definition of AI can vary hugely. For instance, Fei-Fei Li, Co-Director of the Stanford Human-Centered AI Institute, defines AI as a field of study focused on creating systems that can perform tasks that would typically require human intelligence. She emphasises the importance of technology being designed to align with human values and ethical standards. Microsoft CEO Satya Nadella defines AI as an augmentation of human intelligence that can empower people to achieve more. He focuses on AI as a tool that complements human capabilities, fostering creativity and productivity.

For the purposes of this book, we will use a very simple interpretation of AI as follows: 'Computer- or machine-based systems that can learn, adapt and perform tasks that typically require human intelligence in order to enhance decision-making.' We will categorise the potential of AI to provide enhancements in the following four areas: learning, reasoning/decision-making, self-correction and creativity.

Considering these areas as distinct opportunities helps provide context for AI's benefits and allows us to assess its risks. For example, using AI to learn from past financial patterns or in people-focused disciplines is generally less risky than using it to make creative decisions about a company's brand imagery. Despite some overlap, these areas are distinct enough for organisations to separately evaluate the pros and cons of each.

Reasoning, Decision-Making, and Ethical Choices

Organisational leaders should be looking to consider the impact of AI in the workplace through the physical manifestation of the tools and applications that can enhance labour efforts, as well as the consideration for the impact AI can have on decision-making. This raises a philosophical challenge relevant to AI in general, and certainly pertinent to the adoption of AI within organisations: The information on which AI is trained, how the guardrails are set and the way in which AI is enabled to act and reason are all human-set limitations; consequently, they all come with a risk of bias.

AI systems operate by analysing vast quantities of data and applying algorithms to recognise patterns, forecast outcomes and make decisions. Generally, humans monitor these systems, promoting beneficial behaviours and discouraging those that are not advantageous.

AI will usher in a time where we will make decisions in three primary ways:

(i)   By humans (which is familiar)
(ii) By machines (which is becoming familiar)
(iii) By collaboration between humans and machines (which is not only unfamiliar, but unprecedented – see Kissinger et al 2022).

Neither humans nor machines are devoid of bias. Human bias has been a topic of research in psychology for many years. Any bias in AI could be viewed as a consequence of human bias: The data on which a machine learns, the initial programming, the data left out from informing an AI system and the way measurement of the system is done are human-led processes. The process of learning is broadened from moving from labelled to unlabelled data to discovering patterns and relationships within it. Reinforcement-based learning further trains the model: Additional steps can assist in dealing with the bias introduced to the process by a human – but these steps can only be carried out if the human is aware of their own biases! A closer look at the challenges of bias will come in Chapter 10.

Applying AI: The Dilemma of Differing Opinions

There are many philosophical and ethical decisions to be made around AI. As we will see later in the book, there can be significant reputational risk if the approach to AI deployment is not carefully considered and executed. The late Stanford Professor John McCarthy, one of the founders of the discipline of AI, described the link between philosophy and Artificial Intelligence thus: 'Artificial intelligence (AI) has closer scientific connections with philosophy than do other sciences, because AI shares many concepts with philosophy, e.g. action, consciousness, epistemology (what it is sensible to say about the world), and even free will' (2006).

AI presents a number of new challenges for which our society and organisations are currently unprepared. Navigating these topics requires leaders of organisations to bring together a number of different skills across their companies. On the one hand, there are areas requiring the skills arising from the experience of living through many years of similar issues – particularly ethical challenges, compliance issues and PR disasters. The wisdom brought to the workplace by members of the workforce sporting a few grey hairs needs to be harnessed here.

At the same time, the youngest members of the workforce are the ones who best understand AI utilisation. They were born in an era where they have always had the full force of the internet at their fingertips, and have recently used GPTs as a 'learning buddy' to help them complete work more efficiently. Having used AI either to increase the efficiency of their daily workplace tasks, or to enhance their learning at university before entering the workplace, these newest entrants' knowledge of how to apply AI is significantly ahead of their older colleagues.

These two different groups have very different views of the world, the workplace, what they expect from a job and how management should be conducted. They may also have conflicting opinions on the philosophical topics of bias in AI and of the organisation's responsibility to mitigate it. The successful implementation of AI will be by leaders who can understand the differences between these two essential groups and bring them together to work in harmony.

AI's Capabilities and Limits

Whilst the potential of AI is incredibly high, it can be over-blown in terms of what it is actually currently capable of. Most of all, current AI models can get seemingly simple things wrong. Whilst the speed of development in AI is making even Moore's law look pedestrian in comparison, chip availability and the sheer computing power and energy required to run AI are likely to slow any potential for computers to become 'self-aware' for a very long time. We are at a stage where AI is largely reactive: good for classification and pattern recognition tasks, great for a scenario where all parameters are known, but incapable of dealing with imperfect information.

In some fields, AI is capable of human – or even superhuman – levels of performance. In others, AI makes many basic errors. Not all tasks can be achieved by AI, a point demonstrated well by the AI researcher and author Janelle Shane in her blog 'AI Weirdness', which lays out examples of AI failing to complete what could be considered very basic tasks.[11] Shane's experiments with AI show the present challenges with image genera-tion and labelling in particular, which anyone who has tried to generate a picture of a person and suffered from a returned

---

11   For instance, efforts in labelling animals in Dall-E3 yielded some totally wrong answers (see https://www.aiweirdness.com/learn-the-mammals-with-dall-e3/). Shane's book, *You look like a thing and I love you: How artificial intelligence works and why it's making the world a weirder place* (2009), is a great (and humorous) way to learn about AI's capabilities – and failings!

image with additional limbs or unconvincing hands can attest. A great demonstration of a failure of current Generative AI is the exercise Shane ran when asking AI to design the ultimate pickup lines, which resulted in the recommendation that became the title of her book: 'You Look Like A Thing And I Love You'.

Some problems, both for machine programmers and for humans interacting with each other in a workplace, were tough to crack before we had access to large AI models and huge sets of data. There is a danger in applying AI to all situations, especially those that just require a bit of common sense – a point Shane makes well, and one that leaders would do well to heed.

How AI is Ushering in a New Phase

Building on the digital and technological revolutions that we have witnessed over the past 60 years, AI is heralding the entry to the next phase of the Industrial Revolution. This phase, like previous phases, will change the way that humans create, exchange and distribute value. We have had centuries of experience of using machines to make manual tasks easier, in some cases replacing manual labour altogether. The opportunity facing us in organisations is developing uses for AI to enhance our labour, as we have previously done with machines.

The shift to digital is within the memory of many current business leaders. Creating efficiencies via this digital transition required significant change through entire businesses. We are used to hearing the term 'digital transformation', which encompassed several stages. The process of *digitisation* – converting information from formats such as paper and analogue audio into digital files – was the initial stage for many. However, this was just one stage: Having information in a digital form is only useful if processes can be put in place that enable these formats to improve company activities. *Digitalisation* involved using new technology and building new processes amongst working teams to realise the efficiencies that digitisation made possible, which could only be done by changing the previous habits of working teams.

For many organisations, the digital transformation was far from plain sailing. Transformation was slow, a lot of money and resources were wasted, and there are still companies today that have failed to transform.

As leaders looking to draw lessons from this period should note, this failing was largely due to companies failing to focus on the people in the organisation and help them make the required changes to enter this brave new era. In our current organisations, the smartest digital products or processes are irrelevant without focusing on the people who need to adapt to them.

Summary

For leaders of teams and organisations, AI is an opportunity that is not without hurdles. As more generative AI tools are launched and greater large language models are created, even more potential is available to companies. AI creates a conundrum in the boardroom: Leaders know that they need to tread carefully, but also feel that they need to move *fast*. A 2023 survey conducted by IBM in collaboration with Oxford Economics revealed that 60% of organisations are not developing 'a consistent, enterprise-wide approach to generative AI'.

Whilst some of AI's output may appear somewhat magical, it has many limitations which need to be well understood in order to grasp the advantages it can bring. Alongside its great possibilities, AI poses many challenges to businesses. Strategically allocating resources and building capabilities around AI could enhance competitive positioning while unlocking significant value. Conversely, misjudgements in this area have the potential to expose organisations to data privacy vulnerabilities, as well as legal and reputational repercussions.

It should be remembered, however, that with AI, we're making *tools*, not colleagues. The collaboration possibilities that exist and the enhancements that can be made are not possible without human instruction and intervention – both of which require a focus on people.

As we go through the chapters in the book, we will look at how leaders need to navigate the opportunity not by looking at the physical system of AI itself, but rather at the skills of the people in their organisations who can help them navigate this great opportunity.

So far, many people's conscious interaction with AI in the workplace may be limited to conversations with a GPT such as ChatGPT; however, AI's applications are far broader and the possibilities of what it can do for an organisation are far greater. Leaders must learn about this technology's applications within their organisation, but also how to apply an appropriate leadership style for the AI age. Our next chapter lays out how business leadership has evolved, as well as the challenges for the modern leader.

GPT Prompts

- Briefly explain what it means when an AI model 'hallucinates'—including why this happens from a technical standpoint. Then, share three real or widely reported examples where AI produced clearly incorrect or amusing responses. Where possible, highlight what these examples reveal about the current limits of AI in business or leadership contexts.

> • Provide cross-industry examples of how AI is currently being used to support—but not replace—human decision-making. Focus on scenarios where AI contributes insights, probabilities, or predictions, while humans retain accountability. Include examples from at least three sectors (e.g., healthcare, finance, manufacturing, or marketing), with a brief explanation of the value AI adds in each case.

What Questions Should a Leader Be Asking?

• Where is AI already being used by my organisation?

• Who is currently responsible for the decisions made by AI within our systems and processes?

• How transparent are the AI systems we currently use, and can we explain any decisions made by AI to our stakeholders?

• What is our organisation's definition of AI, and how does this differ from how other business leaders, investors, regulators or others define AI?

• How can I provide clear and transparent communication about the role of AI in our organisation and its impact on the workforce to alleviate any anxieties or misconceptions?

4

MODERN LEADERSHIP

The Continual Evolution

*"Isn't it funny how day by day nothing changes,
but when we look back everything is different."*

– C. S. Lewis

T he oldest-known business in the world still in operation is the Japanese company Kongō Gumi Co Ltd, reportedly incorporated in 578 AD.[12] Within the world's oldest businesses, 15 are over a thousand years old and almost 250 are more than 500 years old. From construction to hotels to banks to transport to food and, of course, brewing beer and making wine, these companies appear to be very old – older, in fact, than the modern constructs of some nations.

---

12   Operating as a subsidiary of Takamatsu Corporation since 2005.

FIGURE 2 – OLDEST OPERATING BUSINESSES IN THE WORLD
BY COUNTRY

Business Name	Country	Year Opened
Kongō Gumi	Japan	578
St. Peter Stiftskulinarium	Austria	803
Staffelter Hof Winery	Germany	862
Monnaie de Paris	France	864
Sean's Bar	Ireland	900
Marinelli Bell Foundry	Italy	1040
Affligem Brewery	Belgium	1074
Munke Mølle	Denmark	1135
Aberdeen Harbour Board	Scotland	1136
Ma Yu Ching's Bucket Chicken House	China	1153

As we learn more about how people interact with each other within organisations, leadership styles have evolved. This chapter lays out how the leadership of organisations has evolved over time and how it will further adapt to the changes that AI will bring to the workplace.

The idea of a business operating as a collective of individuals in a social group is actually a relatively new idea when compared to our evolutionary process. As we operate in our working environments daily, we can sometimes forget the fact that what we see as 'modern humanity' is the result of millions of years of evolution. Through this evolution, our ancestors figured out that, as a species, we are better off in social groups. These social groups range from those small in number, such as families, to larger groups, such as tribes or nations. Our organisations, small or large, all feature the same principles of group membership as in any social group: Our acceptance by and participation in the group are dominated by complex psychological factors.

The challenge of leading this social group has changed little from when the first business incorporations took place. There is the initial requirement of understanding the motivation of individuals, then the practice of running the organisation: acquiring and using resources, bringing the group together behind a common purpose, providing direction, navigating complexity, evaluating approaches, measuring, adjusting, refining, and re-executing.

Over the past few centuries, we have seen several seismic shifts in how industry operates. Each of these shifts has been a result of technological development. From early mechanisation to the impact of the production line through to the computer revolution and the modern digital and data age, each development impacted the way business was conducted, created opportunities and required leaders to navigate change.

The most significant development of recent times is the opportunity brought by Artificial Intelligence. As we've seen in the last chapter, AI isn't actually that new – but it *is* becoming much more accessible, starting to be used by people inside many businesses to create efficiencies; therefore, leaders must now consider their approach to leadership in the age of this latest technology.

How Leadership has Evolved

The subject of leadership was a topic for philosophers rather than psychologists up until the end of the First Industrial Revolution. Since the mid-19th century, psychologists have looked to make sense of the relationship between leaders and those in the groups they led. There are many books dedicated to leadership, as you'll be aware. This isn't intended to be a book that describes the different theories in any great depth. We will touch on some relevant leadership theories as they relate to leading in the workplace today. (A broad guide is available in Figure 3 following.) These are useful to understand, as they pertain to the modern workplace and AI. For further information on any of these, please look to the GPT that accompanies the book via the website (www.onv. ai), your favourite search engine or your local bookshop.

Over the past decades, leadership thinking has evolved from being dominated by theories based on the leader themselves. There are many different leadership theories from the past century. Some examples of the broad ones are:

- 'Great Man' theory (also known as Trait theory): The notion that leaders are born, not made, having somehow been blessed with a genetic ability to lead others.
- Situational theory, which contends that how a leader interacts with the people they lead should be adjusted to their specific context, situation or environment.
- This theory was enhanced a step further with Contingency theory – the idea that the leader's behaviour and traits

need to be considered in conjunction with the specific situation to decide on the most effective leadership approach.

The list of management theories goes on, encompassing lateral leadership, quiet leadership, integrated psychological theory, and even chaos theory. Each has its own merits, with dedicated books emerging as more research is conducted.

Some theories seem to be *en vogue* for a particular period of time, but two theories that have dominated modern leadership thinking in recent years concern transactional and transformational leadership.[13] In transactional – or exchange – leadership, the leader motivates individuals via systems of reward/punishment for performing/not performing set roles.

Transformational leadership theory sees leaders 'transform' followers through inspiration and creating a sense of belonging to a group where rules and regulations can be shaped far more by group norms.

The evolution of leadership styles is in part due to the changing environment, and changing people, within the workforce. At the end of the 19th century, it was the norm of society to be strongly hierarchical and for people to consider themselves so lucky to have a job that they would put up with any working conditions. The same could not be said of the average worker by the end of the 20th century, however.

---

13   See the work of Max Weber, James Downton, James MacGregor Burns and Bernard Bass.

FIGURE 3 – HISTORY OF LEADERSHIP THEORIES

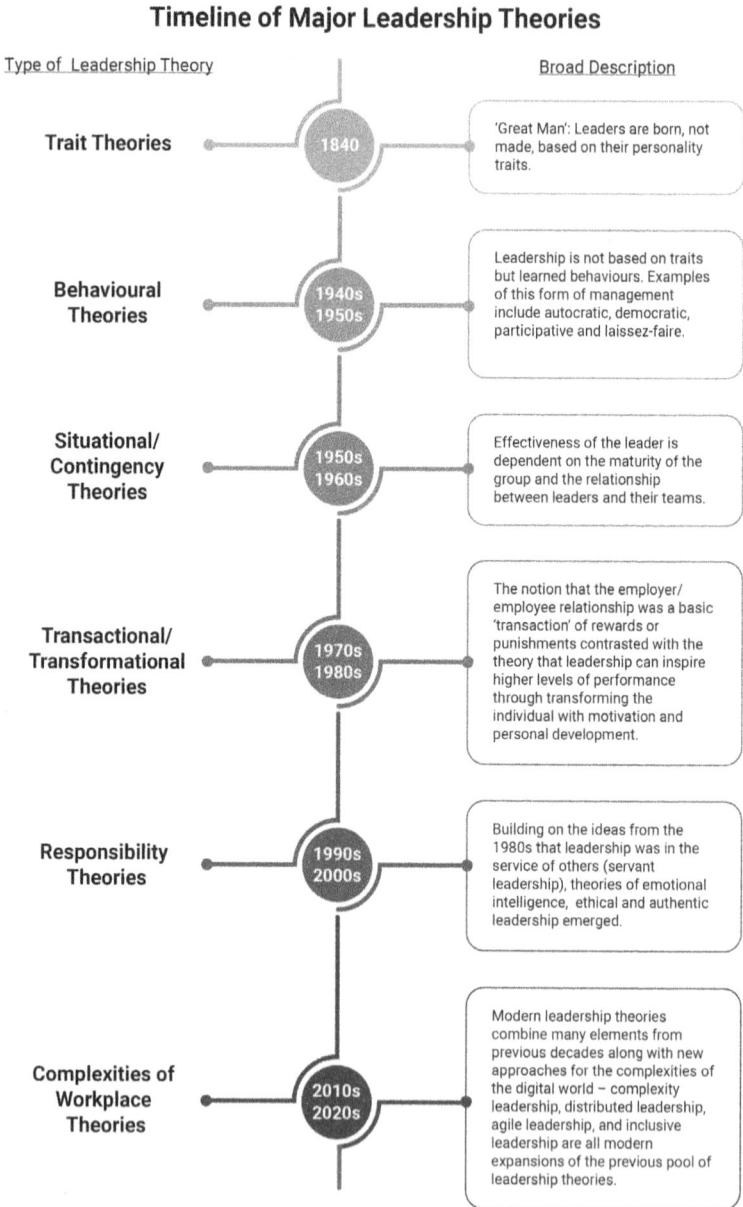

Timeline of Major Leadership Theories

Type of Leadership Theory		Broad Description
Trait Theories	1840	'Great Man': Leaders are born, not made, based on their personality traits.
Behavioural Theories	1940s 1950s	Leadership is not based on traits but learned behaviours. Examples of this form of management include autocratic, democratic, participative and laissez-faire.
Situational/ Contingency Theories	1950s 1960s	Effectiveness of the leader is dependent on the maturity of the group and the relationship between leaders and their teams.
Transactional/ Transformational Theories	1970s 1980s	The notion that the employer/ employee relationship was a basic 'transaction' of rewards or punishments contrasted with the theory that leadership can inspire higher levels of performance through transforming the individual with motivation and personal development.
Responsibility Theories	1990s 2000s	Building on the ideas from the 1980s that leadership was in the service of others (servant leadership), theories of emotional intelligence, ethical and authentic leadership emerged.
Complexities of Workplace Theories	2010s 2020s	Modern leadership theories combine many elements from previous decades along with new approaches for the complexities of the digital world – complexity leadership, distributed leadership, agile leadership, and inclusive leadership are all modern expansions of the previous pool of leadership theories.

Among the common methods currently talked about are Authentic and Servant leadership.[14] Authentic leadership, as the name suggests, involves leaders operating in line with their own values and beliefs, inspiring others through their ethical behaviour. Servant leadership sees the leader putting the needs of their team above their own self-interest, playing a supportive more than a directive role in helping the team achieve its goals. Whilst new terms abound in leadership research, one factor of the successful modern leader is that they are commonly highly emotionally intelligent, understanding that motivating individuals in the modern age needs to be performed by understanding people rather than relying on old transactional methods.

Managing Modern Teams

How an individual interacts with these systems of leadership is highly dependent on their expectations, which are built through experiences and highly influenced by the environment in which they spent their formative years. The advent of AI as a new workplace tool doesn't change the workplace expectations of the individuals in your team. Within the current workforce, there are groups with differing life experiences

---

14   For more reading on the theories outlined here, see the work of Thomas Carlyle, Kurt Letwin, Paul Hersey and Ken Blanchard, Max Weber, James McGregor Burns, Robert House, Bernard Bass, Daniel Goleman, Mary Uhi-Bien, James Spillane, and Alma Harris.

and skills, which the leader must coordinate into one harmonious and high-performing team.

Have you ever heard the popular leadership analogy of the leader as the conductor of an orchestra, expertly coordinating the different instruments into one beautiful sound? The reality of this analogy of modern leadership, as described to me by one CEO, is not that they stand at the front, uniting a combination of different skills to play in unison. The reality is more akin to what would happen if the violinists had decided they wanted to play the cellos instead, the trumpet section had forgotten their sheet music, the percussion section wanted to play an entirely different piece of music altogether, and the oboes were more concerned with the correct height of everyone's music stand before anything can start.

Organisations, like all social groups, operate in a hierarchy. The traditional hierarchy of organisations typically has an age bias. The successful adoption of AI is heavily dependent on bringing together skills from different age groups in the organisation. Information flow is essential to make the most of the skills that exist within these different groups. The unification of the new (younger) entrants and established (older) team members means understanding their needs and appropriately engaging with different groups.

These groups may differ in the following ways:

(i) Learning Styles

My education, like the generation before me and many of the generation after, was a process of 'learning by rote'. The teacher stood at the front of the class, chalk and blackboard were the tools of the trade, and we memorised many things by repetition – plus the fear of punishment from a metal ruler for getting it wrong. (Such was the joy of the educational system in England in the '70s and early '80s.)

The classroom experience of the newest entrants to the workplace was very different. The use of technology in the classroom has a lower focus on the 'static' approach, more on the experiential.

(ii) Learning Types

In the 1960s and 1970s, far fewer people than today went on to complete university. Just 4% of school leavers in the UK did so, rising to around 14% by the end of the 1970s (Lightfoot 2016). That figure now stands at more than 40% of young people undertaking degrees. The types of degrees people are choosing are also changing. In the US, the number of people completing degrees in business and health practices has increased, whilst the numbers pursuing degrees in education, social sciences and history has diminished.[15]

(iii)   **Places of Work**

In recent years, employees have increasingly demanded flexibility from employers in terms of work location. Deloitte's annual survey of Millennials and GenZers is an excellent place for leaders to look for trends and attitudes – nearly 23,000 people surveyed in 2024 across 44 countries. Despite a push from many corporations to 'return to office', there is still a huge number of GenZs and Millennials working remotely – 50% of GenZs and 46% of millennials surveyed have a hybrid or remote working pattern. (This is lower than the 65% of GenZs and 64% of Millennials who want one.) The 2023 survey showed that 77% of GenZs and 75% of Millennials currently in remote or hybrid roles would consider looking for a new role if their employer expected them to be full-time in the office; however, in the 2024 survey only 13% of GenZs and 11% of Millennials said that they started looking for a new role as a result. Ignoring the desires of this section of the workforce could lead to talent attraction and retention issues. The leader needs to strike a balance between younger generations' needs and older

---

15   U.S. Department of Education, National Center for Education Statistics, Higher Education General Information Survey (HEGIS), 'Degrees and Other Formal Awards Conferred' surveys, 1970–71 through 1985–86; Integrated Postsecondary Education Data System (IPEDS), 'Completions Survey' (IPEDS-C:91–99); and IPEDS Fall 2000 through Fall 2020, Completions component. (This table was prepared September 2021.)

generations' notion of the higher productivity and collaborative benefits of being in an office together.

(iv) Styles of Workplace Engagement

One of my coaching clients, Sarah, is an executive who entered the workplace at a time when memos and faxes were used to communicate info. Agendas were circulated five days in advance of a meeting, with minuted notes circulated afterwards. She made the shift easily to email in the '90s, where long email chains (including extensive cc lists) replaced memos.

When Sarah came to work in the tech world, she found her teams preferred daily stand-ups with no previously agreed agenda, and workplace productivity tools had even replaced email. She was frustrated by the changes made in collaborative documents with no versioning system, making it challenging to track the changes made. The youngest team members were using Slack, Asana and Google Docs, while the older ones used email, Microsoft Word and Excel.

Sarah had to make the decision to put aside her own preferences and develop communication and working tools that were most appropriate for getting things done. This meant selecting tools that were effective not just for ease of communication but accuracy as well. Legal documents had to use methods of shared editing that other teams could understand. Financial documents had to have a versioning system in order for everyone to understand that if a change was made, it was known about and could be easily traced. She involved her team in discussions about what tools would offer this capability, balanced this against the individual preferences that many had (based on their prior experiences), and avoided silos by making information available to all. This required some change management for members of the team – but ensured that information was organised in a way that was easy to find and, vitally, accurate. Sarah's decisions regarding the tools selected were not popular with every member of the group, but even the discussion on which tools to use and why brought her team closer together in understanding their differences.

(v) Feedback

In one of my earliest jobs, feedback was reserved for the situations where I did something wrong. I knew that I was doing well if I kept my job, didn't get shouted at by any of the senior team and got a bonus and a pay rise each year. I remember this annual 'review' where I would be told something along the lines of, 'You've had a good year, we're really pleased with you, so we're going to give you X'. Literally a five-minute process. Times have changed – and fortunately, leaders understand the benefit of feedback. The more astute engage both themselves and their leadership teams in a 360-degree process of feedback, upwards as well as downwards. The younger generations both expect and are receptive to regular feedback.

(vi) Opinions on Engaging with AI

Nearly 20 years ago, I was at my friend Steve's house for dinner. 'Watch this,' he said to me, and asked his nine-year-old son to turn the light off. He then asked him to turn it back on. 'Did you see that??!' he exclaimed. In my rush to presume he was slightly mad for celebrating the fact that his son could control a light switch, I'd overlooked a subtle detail that he explained: 'He uses his thumb to switch the light on and off! You or I would always use a finger.' (He was right.) 'Jake, on the other hand, has grown up playing games – the thumb he uses for control in those games has become his primary digit!'

These changes in how we operate physically stick with us as we grow older. Relying on a search engine to do your homework will lead to you leaning on it in the workplace. Using a GPT as a form of information in college will lead to the same outcome. The likelihood of going past the first page on a search engine is now very low, such is it trusted. The same will happen with GPTs in the workplace, and this can lead to a conflict of opinion: The young may have grown to trust it, as they have been using it for a while, while the older generation may distrust it until it has been proven to be accurate. The answer to which of these approaches is appropriate should be decided through a collaborative process involving the opinions of all parties, taking into account each task's need for accuracy.

(vii) AI Ethics – How Do Values Differ?

AI faces ethical challenges. The data used for training an AI system must be inclusive and diverse. Early stories of machine learning gone wrong due to the data used to train the system grabbed headlines – yet the cause was that the machine learning did exactly what was asked of it, but the *inputs* were biased. Amazon's resume screener (Dastin 2018) that was biased against women, the COMPAS algorithm used to predict re-offending (Larson 2016) and the racial bias found in a health-care risk assessment algorithm (Vartan 2019) are all examples of how bias can creep into AI. But other ethical considerations also exist – for instance, the company's mitigation of the environmental impact of the increased processing power required for complex AI processes, or the company's responsibility to employees whose roles are replaced by AI (including retraining those most at risk).

With younger generations putting a much higher emphasis on the company responsibility for inclusivity and social responsibility, falling short in this area could mean more than public reputational damage – it could lead to staff leaving and a failure to attract future talent.

The ethics of AI is possibly the most important topic for societies to reach consensus on. Who should provide the oversight to ensure that AI complies with societal values, the role of governmental bodies and how far they should influence organisations implementing AI is a debate that will continue for some time.

Summary

This chapter has given a high-level view of how leadership has changed over time and how it must adapt now to the changes that AI will bring to the workplace. At the heart of this is an understanding of the needs, hopes, dreams and desires of the people you lead. As we will unpack in later chapters, the era in which people were born and the influences on their life in their formative years have a huge role to play in the values and norms that they bring to the workplace.

GPT Prompts

- Summarise the origins and core principles of both Transactional and Transformational leadership models. Explain how each approach emerged historically, and outline the key differences in how they influence motivation, change management, and team performance. Include insight on when each style may be most effective in a modern organisational context.

- Build a list of bullet points for a middle manager on what someone under 25 expects from the workplace versus someone over 55.

What Questions Should a Leader Be Asking?

- How has my leadership style evolved over time, and are the approaches I employ in leadership appropriate for leading the groups I need in the AI age?

- Does my leadership style deliver the connection I need with different age groups in the company?

- Is my communication approach within the company appropriate for pre-digital and post-digital generations?

- What strategies can I implement to foster open and transparent communication, both within the leadership team and across the organisation?

- How could I build and nurture strong relationships with team members, fostering trust and collaboration in a virtual or geographically dispersed setting?

5

THE GREAT EFFICIENCY ILLUSION

How did we become so dysfunctional in the data-rich era?

"The definition of insanity is doing the same thing again and again and expecting different results."

– **NOT** ALBERT EINSTEIN[16]

Many digital transformation projects fail. Leaders must use what has been learned from these failures to avoid the same mistakes when transforming in response to AI. In some cases leaders may not be able to see the reasons clearly, as they are not close enough to the

---

16   Al-Anon? Narcotics Anonymous? Max Nordau? George Bernard Shaw? Samuel Beckett? George A. Kelly? Rita Mae Brown? John Larroquette? Jessie Potter? Werner Erhard? 2010, *The Ultimate Quotable Einstein*, Edited by Alice Calaprice, Section: 'Misattributed to Einstein', p. 474. Princeton University Press. (Verified on paper.)

company's transformation efforts. In the worst cases, they may be deluding themselves that the transformation work has been as successful as it could have been. This chapter is designed to unpack some of the reasons for these failures, and to help leaders identify where they are currently with their organisation's development around data.

In 2016, the consultancy firm McKinsey released an article designed to help leaders with digital transformation efforts. In this article, they shared that 70% of complex large-scale programs don't reach their stated goals, arguing for the creation of a 'performance infrastructure' (Bucy 2016).

Adoption of AI requires further business transformation. Whilst there is strong evidence to show the value of digital transformation,[17] plenty of money has still been wasted by investing in failed projects.

The term 'digital transformation' is a blanket term that covers many parts of a multifaceted process that integrates digital technologies into all areas of a business. Over time, the focus on the enabling factors of a digital transformation can change. Throughout the past decade, extensive focus has been on cloud-based solutions and data infrastructure. The shift to AI for some companies requires building on top of past

---

17   The calculations in the 2023 book *Rewired* by Eric Lamarre, Kate Smaje, and Rodney W. Zemmel, using McKinsey's Finalta benchmark, offers a good look at the financial returns of digital leaders versus digital laggards.

THE GREAT EFFICIENCY ILLUSION

developments, while for others it will require the development of new infrastructure.

Spend on digital transformation technologies and services is accelerating. This figure has continued to accelerate since hitting a global annual spend of $1 trn in 2018, and is forecast to grow to $3.9 trn by 2027 (Sherif 2024). Years of observing the failures in digital transformation projects – in spite of the vast sums of money invested – have made me realise that there is something fundamental missing when most transformations have been undertaken, which must be rectified with the adoption of AI.

Some of the top reasons for the failure of a digital transformation are a lack of alignment with business outcomes, a lack of awareness within the organisation and a tendency of some to focus on the 'shiny new object' – blinded by the apparent glitz of a digital tool that doesn't have a business case for adoption.[18] There are many more things that make digital transformation difficult and contribute to the failure figure, but ultimately it is often down to focusing on the wrong element in the transformation.

The order of importance of the different elements needed in digital transformation vary by business (and which seller of a tool or process promising your desired transformation you

---

18   For a more comprehensive list, see the Forbes Coaches Council article '12 Reasons Why Your Digital Transformation Will Fail'.

talk to), but most seem to agree on at least the following five elements being present in order to execute a successful digital transformation:

1. Access to the right tools/technology infrastructure
2. The right processes to follow
3. Executive buy-in/leadership
4. The right people within the right organisational structure
5. The right approach to implementing change.

Although much of the rhetoric about digital transformation talks about high-level goals and shared vision, the primary focus seems to have been on the first two elements – the right products and the right processes. The digital transformation industry is certainly full of products and people who would like us to adopt a certain process – paying handsomely for the dashboards to go along with it.

A great way of visualising how innovation spreads among people is the 'Diffusion of Innovation' theory developed in 1962 by Everett Rogers (1995, p. 12). Many of the terms from his research became commonplace in both business and consumer language, used to describe the position of either a company or an individual with respect to the adoption of a new idea or product.

Rogers' research showed that adoption wasn't even across the whole group, Instead, some are more likely to adopt new ideas or products than others, at different stages. He identified five adopter categories, which could be broadly split (dependent on the product or idea and social group) into the categories of innovators, early adopters, early majority, late majority and laggards.

FIGURE 4 – EVERETT ROGERS' DIFFUSION OF INNOVATION MODEL

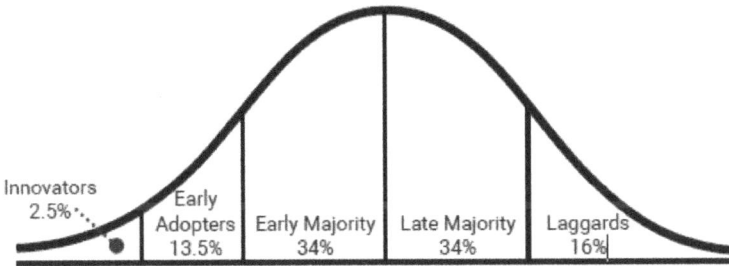

It may be useful to consider not just where your company fits across this dimension, but where the *individuals* in the company sit. Other factors of course influence these stages – the awareness of the need for the product or idea, the decision to adopt/reject the innovation, the trialling of the innovation and its continued use all come into play in each of the categories.

As I've observed in organisations I've worked with, the same is often true of how innovations are adopted. Leaders need to be aware of the factors influencing the adoption, then use these to position a new innovation:

(i)   How much better is the innovation than what it is replacing?
(ii) How does the innovation match the needs and values of those you want to adopt it?
(iii) How difficult is the innovation to adopt /understand/use?

(iv) How easy is it to try out the innovation before making a commitment to adopt it?

(v) Does the innovation provide tangible results?

As leaders negotiate these factors, a subtext that can occur within some organisations is the resistance to innovation. Leaders should bear in mind the question that many of their team may ask themselves (if not aloud): 'What's in it for me?' or 'What does this innovation give me that I don't want or won't accept?'. These underlying questions need to be addressed to elicit change. Self-preservation is a common reaction and is often weighted to the short term. Consider within this introduction of innovation what is at stake for them and their futures: Without adaptation to the changing world, people can very easily be left behind in the workplaces of the future.

Consider also how much the different generational groups feel a sense of ownership regarding the change. As I talk to people from my generation (Gen X) and Baby Boomers about the influence of digital technology on their workplace practices, we were influenced primarily by the introduction of technology for working tasks. For GenZ and many Millennials, however, the influence of digital technology had on how they operate was completed before they arrived at the workplace. As a result, they are confronted with processes and working practices that they may not see as optimal.

A Brief History of Digital Development

It is useful at this point to look at both the availability of digital technology and the skills of the people who were asked to change to adopt it. Whilst the origins of the digital era started with the invention of the microchip and semiconductor back in the 1950s, for our purposes of digital transformation, we will look at how our organisations have progressed through the internet and mobile era.

In the year 2000, worldwide internet adoption stood at 414 million people. By 2010, that figure was 2.02 billion (World Bank 2022, UN 2022). Many older adults were first exposed to the internet in the workplace. By 2010, there was a stark difference between the newest and the longest-standing workplace members in terms of overall adoption. In the United States, 92% of 18–29s were using the internet by 2010, compared to 74% of those aged 50–64 (Pew 2021). The younger generations were exposed to this technology at school or university and increasingly at home, as internet connectivity matured from dialup modems into broadband connections.

Even more staggering was the increase in mobile phone adoption. As the first decade of the new century unfolded, mobile phone subscriptions worldwide grew from under a billion to over five billion. Whilst initially the preserve of businesses who could afford both the high entry cost of handset purchase and the subscription charges, more young people adopted the technology as phones and subscriptions came down in cost.

FIGURE 5

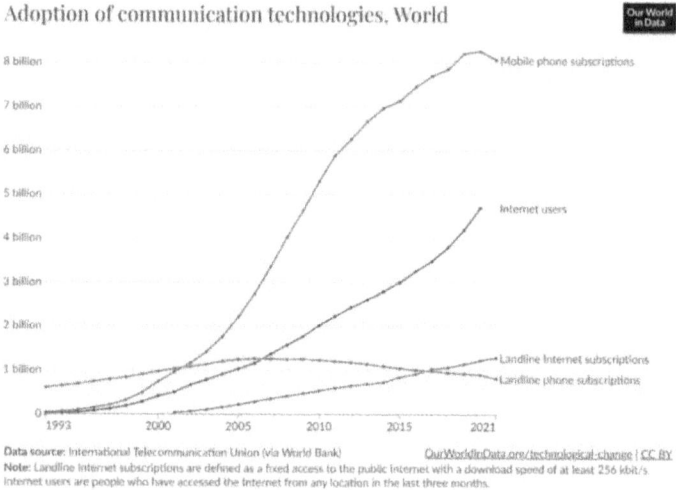

Adoption of communication technologies, World

International Telecommunication Union (via World Bank); Gapminder (2019); UN (2022); HYDE (2017); Gapminder (Systema Globalis) – processed by Our World in Data. 'Internet users' [dataset]. International Telecommunication Union (via World Bank); Gapminder (2019); UN (2022); HYDE (2017); Gapminder (Systema Globalis) [original data].

For those in their teenage years during the early 1990s or before, arranging to meet friends involved a rigid time and place: If your friends didn't show up, there was very little ability to find out if they had forgotten, were lost or worse. Mobile phone adoption solved many issues, not least the ability to let a friend know of a delay in arrival.

As I entered an advertising agency in 1993, computers were almost all 'desktop'. Communication of advertising schedules of activity was done by fax machine. Important documents or visuals for campaigns were couriered on pushbikes/motorbikes around cities. Printing of materials that were used in evaluation of campaigns was such a slow process that it was scheduled to be done overnight, and printed results came on long chains of concertinaed paper with perforated sides. One of the lowliest jobs, bestowed on new graduates, was to pick up these overnight reports, then separate them to place on the right desk for the buyer who ordered the ad campaign (making absolutely sure you removed all the perforated edges). Not the most exciting task in a team of 80 people and reports of several hundred pages every day!

By the year 2000, communication was performed mostly on email. Personal mobile phones were text-based and Blackberrys were all the rage, enabling people to send and receive email on the go. Around this time, instant messaging services on desktop computers such as MSN Messenger and AIM started to be used for quick communication – mostly brought in by the younger entrants to the workplace. The growth of these services also seemed to reduce, if not eradicate, the calamities of people mistakenly 'replying all' to a company-wide email, or trying to recall an email that was erroneous. Receiving an email headed 'John would like to recall the email' just seems to pique people's curiosity more...

The adoption by age of digital technologies is highly relevant to digital transformation efforts. Our colleagues who went to school in the late 1990s and early 2000s have never known a time without being able to be constantly connected to others, or without having access to the internet at school, home or a library – a level and ease of access to information unprecedented in previous decades. For people entering the workplace after growing up with the internet or a mobile phone, the norm was of rapid access to information and quick connection to others. Even the brevity of messages they were accustomed to through instant messaging services was brought into the workplace.

The workplace 'digital transformations' featured changes within companies centred around the internet, mobile communication and 'always-on' connectivity. We saw new skills being brought into the workplace and a new wealth of data that could be captured, stored and analysed. Customer and employee experience could be enhanced by the collection and analysis of this data. In 2006, the phrase 'Data is the new oil' was coined by Clive Humby, co-founder of the data agency Dunnhumby.

The promise was that this data would facilitate the finding of insights, lead to greater automation and ensure that machine learning could be run on the company's vast troves of data to create new customer offerings.

The reality is that the use of this data, much like the digital transformation process, has been less than plain sailing. There are groups clearly in line with the diffusion of innovation theory – some of which were brought about by availability and exposure to technology simply reflecting the era in which they were born.

The 4 S's of Data Strategy

My experience over the last decade working with large companies' marketing teams has been that many companies believe they are further ahead than they actually are with the execution of their data ambitions. Common mistakes include a lack of preparation for which use cases the data were supposed to serve, building flawed data 'foundations', holding a random collection of as much data as possible, and trying to build on top of incomplete or inaccurate data.

During my time at Meta, senior data scientist Kate Minogue and I shared our experiences both from working in companies with data, and working with third-party companies trying to help them make sense of their data. This led to Kate's below visualisation of the common stages we found in company data sophistication. It was this work in the Marketing Science space that enabled us to help our clients identify where they were in their development and make changes in order to maximise their use of data.

We looked at the stages we commonly saw amongst companies with data with respect to their methods of organising around data – how data processing and storage was conducted, how data was commonly used and what structure there was for leadership and data teams. We called these four stages 'Struggling, Surviving, Sailing and Soaring', each holding the following characteristics:

Struggling

Characterised by: Frequent data and technology gaps appearing.

Data collection and processing: Excel/spreadsheets as a primary data storage and data sharing tool. Data collection sporadic and unplanned. (We referred to this as the data lake being used like a 'man drawer'.[19])

---

19   The 'man drawer' concept may be specific to some cultures, but it seems to be particular to the male gender. I, and every other man I have ever spoken to, has one of these – a drawer reserved for old remote controls, spare batteries, charger cables, keys to an apartment that we moved out of a decade ago, a screwdriver, defunct mobile phones etc. We throw these all into one drawer just in case we may need them again one day. Should we ever need to retrieve something from it, we spend at

Data usage: Ad hoc/Firefighting. Analysis is as needed and largely reactive. Seen as a necessary evil.

Team structure and leadership: No leadership, direction or expertise. Team structure not yet considered.[20]

Surviving

Characterised by: Starting to invest in data technology and definitions.

Data processing: Databases and even cloud storage may be in use, but they are not the same across all teams. Beginning to formalise and improve data quality and strength of infrastructure.

Data usage: Functional and short-termist. The measures used to assess progress (commonly Key Performance Indicators or 'KPIs') are not aligned, with different departments having conflicting goals. (Sometimes even different teams within the same department have conflicting goals.)

Team structure and leadership: Every team for itself, with tunnel vision. Often there is a need to prove the value of the data

---

least 40 minutes fumbling through the junk and getting distracted by looking for the charger for the old Nokia 3210 for a nostalgic game of snake.

20 In some cases not yet required. We spotted this distinction as important: Some companies rush in to hire data scientists at this stage, believing that this will deal with the problem. Throwing the wrong kind of expertise at the problem leads to false starts, wasted money and dashed dreams.

work being done in order to get more resources/focus. Data strategy driven from the bottom up, with individual teams all planning in isolation.

Recruitment is sporadic and sometimes misguided, with the wrong skills being hired due to lack of foresight/knowledge of different data roles.

Sailing

Characterised by: Building first data and tech strategy. Data value/ ROI has been recognised.

Data processing: Steps taken towards a single data warehouse with strong architecture and dedicated resources. Robust tools and infrastructure in place.

Data usage: Reliable and regular reporting on business health with metrics that are consistent between different teams. Test and learn roadmap for priority data projects.

Team structure and leadership: First structure attempted. Recruiting is done to fill obvious gaps, with the beginnings of discussions of a 'home' for analytic teams – the benefits of centralisation or leaving analysts within each function. At this stage there is quite often a senior sponsor – but not always the right one. There can be either a person 'voluntold' to take this area on, or someone who recognises that taking control of this area could be a path to a greater leadership role in the future. In either case, a lack of skill/experience in this area, or having the wrong level of person who cannot exert influence in the right way, can lead to slipping back to earlier stages of data sophistication.

Another factor observed at this stage is that the tenure of analytics employees can be short due to lack of clear progression or frustration at doing unsatisfactory tasks – this is common with a lack of understanding in the way a person in a data role wants to progress, and the kind of tasks that are interesting to work on.

Soaring

Characterised by: The organisation having an advanced and holistic data strategy. Data strategy in sync/driving company strategy. Data is the core driver of decision-making and innovation. Consistent definitions of success exist regardless of department.

Data processing: Strong data governance and data quality measures. Single unified data warehouse (most often cloud-based, but this can be influenced by location and regulatory factors).

Team structure and leadership: C-level owner with a board seat/board reporting, potentially a Chief Data Officer. Role understands all aspects of data management and the role of different disciplines within the data team.

There is a proven analytics organisational structure that has been decided given the considerations of the benefits of centralised as well as specialised teams. A recruiting plan exists to attract top talent with a sustained career plan for all data roles.

A downloadable 'one-pager' of these stages is available via my website (www.onv.ai).

The main reason getting to the 'soaring' stage seemed to be reserved for so few was that the work is complicated, and that complexity led to relying on specialists. There is a decision we all need to make about being a specialist or a generalist, as summed up in the adage that the trouble with being a specialist is that you end up knowing more and more about less and less until you know absolutely everything about nothing. But unless you are a specialist, you end up knowing less and less about more and more until you know absolutely nothing about everything (Abercrombie 1915).

People Drive Transformation

Data projects are just one element of the transformation process that are considered as organisations make an overall digital transformation. This work involves far more than just a group of specialists. It is an undertaking that must involve the entire organisation, and must pay close attention to having people at the heart of all initiatives. Reliance on a top-down process without understanding and allowing for the human elements of the change process will mean your transformation project stands a very small chance of success.

In the last few years, we are fortunately hearing far more of the learnings that have come from focusing more on process and product and not enough on the people involved.

In a survey conducted by the *Harvard Business Review* in 2022, the top reasons given for digital transformation failing were much less about relying solely on a particular process or type of technology, and more about a failure to acknowledge the role of people at the heart of the change:

Reasons cited for failure were due to a lack of:

1. Embracing digital transformation across our entire organisation
2. Aligning digital transformation with business objectives/ KPIs
3. Effectively allocating resources to the right transformational areas
4. Finding/securing top talent to support our digital initiatives
5. Creating/supporting a culture of continuous learning.

Summary

Digital transformation in the age of AI will encounter the same issues as previous digital transformation efforts if we don't learn from them. For leaders, this means looking at where past changes floundered, but also the influence of the different blend of people we have today. Our most recent efforts in data integration give us very relevant learnings for AI, as they are deeply entwined with it.

People must be kept at the core, and in doing so, leaders must acknowledge that they each come with their own needs and agendas – and that this often differs by generational group. Whilst the newer entrants to the workplace in 2024 have more experience operating with AI, they are unlikely to be ready for dealing with the many ethical and risk-related questions that the company has not faced so far. These are certainly best fielded with the experience of those who have successfully operated within ethics and risk management parameters.

It will require a careful approach to the measures that are used to assess progress (the Key Performance Indicators/KPIs) so that efforts can be managed without silos emerging. This can only be successful if departmental leads unify at the most senior level to ensure the change is successful. It will need upskilling and reskilling not just of practitioners, but also managers who will need to develop new skills for managing in the AI age.

Above all else, it will require careful management of the human element of the change across all parts of the organisation. How this group will form is not something that can be forced – group participation and interaction is a process that can be helped with the understanding of some psychological processes. We will look more closely at the dynamics of how groups are formed in the next chapter.

GPT Prompts

- Analyse how internet usage patterns vary across different age groups, particularly focusing on digital fluency, device preference, and frequency of online engagement. Summarise the key workplace implications of these differences—such as training needs, communication styles, or digital adoption risks—and outline what leaders should consider when designing cross-generational digital strategies.

- As an organisational culture specialist, describe how differences in internet usage by age have impacted workplace practices, culture, or performance.

What Questions Should a Leader Be Asking?

- What strategies has your organisation pursued for upskilling or reskilling employees through digital transformations?

- How do you help people understand what's at stake for them and their futures? How is it different by generation/function/geography?

- Where are your digital transformation efforts relative to the 4 S's of data strategy, and how ready is your team for implementing change management for AI? Would all departments of your organisation agree with this assessment?

- What collaboration is needed between different departments or teams to ensure successful AI implementation?

PART 2

THE PEOPLE WE NEED

6

GROUP FORMATION

What We Can Learn from Dolphins
and Pool Tables

*"I wouldn't want to belong to any club that would
accept me as one of its members."*

– GROUCHO MARX

I n this chapter, we will look at how workplace groups are
formed and the influences that members of the group
bring with them to the workplace.

Leaders in the AI age need to be able to understand both the
motivations of individuals and the values and norms that they
bring into the organisation if they want to influence change
when undergoing a transformation around AI.

When I was in my early twenties, I had the opportunity to
visit a marine research facility where they were observing

the interaction between dolphins and humans. Dolphins are pretty remarkable creatures with some interesting parallels to humans – were you aware, for instance, that they each have signature 'whistles' they use to introduce themselves to new groups, like a first name? (Janik et al 2006). The dolphins in this research program were all wild animals that returned freely to the enclosure where the research was conducted. (The allure of the easy fish that they could earn for interaction was a motivator, I'm sure.) The encounter was one where you would be close to the dolphins, who may interact with you but couldn't be touched in any way. After a morning's briefing we entered the water for a short period, with the researchers monitoring the dolphins' behaviour and reaction to a new group.

We were instructed to swim around the enclosure in a large 'square' rather than swimming straight up to them, as this is the way dolphins will approach a new group. They advised us to bring energy into the initial clockwise swim, and with flippers, mask and snorkel I kicked hard and fast for the first 50 metres or so. It was at this point that a dolphin swam up underneath me, facing me, less than a metre away. This was an incredible moment to be so close to such an amazing creature who was fascinated by me. Forgetting a very important part of the briefing, I reached out to touch the animal. She quickly flipped around in the water and, with a deft flick of her tail, took the mask off my face and snorkel out of my mouth, leaving me floundering and spluttering in the water.

I discovered in the debriefing from the researchers afterwards that this particular dolphin was 'a cheeky teenager' and hung around to have a good look at me spluttering before finding something else to interact with. I'm often reminded of this moment when entering a new social group – there are norms that the group has established (the clockwise swimming direction) as well as a consequence for flouting the social norms (reaching a hand out to a wild dolphin can be seen as a threat).

Becoming a Group Member

Families, tribes, towns and villages, religious groups, sporting groups, educational groups, motorcycle gangs and of course the organisations that we work for are all examples of social groups we belong to. At its most basic, participation in social groups provides protection and the benefit of collective skills. Our ancestors could both defend themselves better against invaders working as a collective and combine resources to hunt, gather and build more effectively.

Our level of participation and motivation for joining these groups varies by individual and the group we are seeking to join. Modern-day participation in a social group provides protection, but there are other benefits, particularly psychological benefits that have been identified as motivators for both joining groups and engaging in ongoing participation.

One theory is that we have a source of both pride and identification in the groups we belong to.[21] This is a powerful concept, and one to be aware of within the workplace, as our need to 'belong' can be powerful if harnessed correctly.

Groups typically have a set of 'rules' for participation. Sometimes these rules are written and explicit – for instance, laws governing expectations of behaviour. Quite often, these rules are implicit but never written down. Sometimes, these unwritten rules can be complex based on group norms and values. For instance, integration based on group norms – growing up a Catholic, knowing when to stand and when to kneel during mass, and whether I should genuflect with my left foot or right foot forward, showed I was a member of the group.[22]

Invisible Rules

An example of unwritten social norms I witnessed was on pool tables in pubs in the UK. These machines would take coins in order to release the balls for a game. The accepted gesture in pubs if there were people playing was to put your coin on the

---

21   See Tajfel and Turner's Social Identity Theory, 1979.
22   The act of right knee to the ground, left foot forward is reserved for worship of God. The act of left knee to the ground was how knights would kneel when getting knighted. Being known as a 'left-footer' as a Catholic, however, is not to do with genuflection but the way the ground was dug in Irish farming. As my Great Aunt explained to me on her farm when I was young in Southern Ireland, using the right foot whilst digging was a sign of a person being a Protestant.

side of the table. If other people had placed coins there already, you took a mental note of how many games were in advance of yours; eventually, when your coin came to the front of the queue, you would challenge the previous winner to a game. If you won that game, you got to play the next person (generally your friends, who would place coins at the same time as you). This was a subtle social norm, but I've never seen it written down anywhere: The rules of pool were often printed and framed on the wall, but there was nothing about how to wait your turn. Flouting this rule by jumping the queue or using someone else's coin could easily cause a fight, as an American friend of mine discovered in a particularly tasty pub in the East End of London once. A large local man leaning over the table explained to my friend in a deep, menacing voice: 'That might be the way things are done over the pond, son – but in here, that's going to cost you a round. For everyone in the bar.' Motivated to stay in that pub, potentially come back again and keep his nose intact, my friend did the smart thing and pulled his wallet out. Years later, he's still warmly received in the bar, where even the bar staff will only refer to him by his nickname, Fosbury.[23]

---

23 Bestowed on him that day for being a queue-jumper, the group gave him the nickname 'Fosbury' – short for Dick Fosbury, a famous American jumper.

What Motivates Workers

Each workplace culture is unique. As we enter a new workplace, we each bring with us our unique blend of motivations, expectations, norms and values. Some of these are based on a combination of our experiences from participation in workplace groups, others from different social groups. Adaptation to some group norms may be easier than others depending on the depth of our relationship with these social values.

Individual motivations to perform also vary from person to person. Psychologists often talk about 'intrinsic' and 'extrinsic' motivations. 'Extrinsic' motivation derives from material benefits – things like salary, promotion, benefits and working conditions. 'Intrinsic' motivation is more connected to self-actualisation – including benefits that come from performing the job itself, such as a sense of achievement.

On the day that Facebook made their Initial Public Offering in 2012, thousands of employees gathered early in the morning in 'Hacker Square' (the central area of the Facebook campus) as the leaders of the company rang the market opening bell to signify the public availability of Facebook shares. On that day, employees were all briefed to avoid conversations with journalists, several of whom were waiting outside the building to try to catch reactions from employees. One unnamed engineer did decide to engage with a journalist, who asked the question:

'What are you going to do with all that money?' His response? 'Socks. I need new socks' (Dembosky 2012).

To presume that someone else would automatically be motivated by the same thing as ourselves would be a mistake. The engineers who are driving AI innovation may have some motivational elements in common with the teams selling the end product, but to presume that, and to try to motivate all team members with purely financial incentives, may miss the opportunity to tap into the motivators that drive exceptional performance.

To effectively manage people and help navigate change within our organisations, the leader needs to understand (i) a person's motivations to work; (ii) their motivations to belong to the group; and (iii) the influences of the individual's values and norms and how they may relate to the organisation. These elements are generally under the surface – they exist in all of our workplace interactions while remaining unspoken.

Above and Below the Surface

Edgar Schein, a social psychologist and MIT professor, laid out a really useful model for thinking about workplace culture. Often described as 'the iceberg model', Schein described three layers that exist in organisational culture (1985). Firstly, there are 'artefacts' and behaviours. These are the visible elements that give an indication of company culture: dress code (indicating formality), open plan or closed offices, and communication methods and decision-making processes. The artefacts are tangible, including the physical structures, organisational language, rituals and ceremonies, and stories and legends that set out what the people in the company believe they collectively stand for.

The next layer of culture is referred to as 'espoused values'. These are often seen in mission or vision statements, but also in the communication by leaders and managers. These are the ambitions and ideals that the organisation has which are expected to guide employees. It is definitely worth noting that in some companies, there can be a gap between these aspirations and the day-to-day reality and practices of people in the organisation.

Under the surface of an organisation, below the daily practices of operation, there are the shared assumptions and shared values that exist as subcultures. These are unwritten and largely unspoken. Over time, people's experiences are shaped by their interactions with others in the company and often informed by how behaviours are rewarded (for instance, how teams should collaborate either within a single team or cross-functionally, or whether individual performance would be rewarded above team collaboration). These underlying assumptions are often deeply engrained, not explicitly talked about and rarely questioned or challenged.

GROUP FORMATION

FIGURE 6

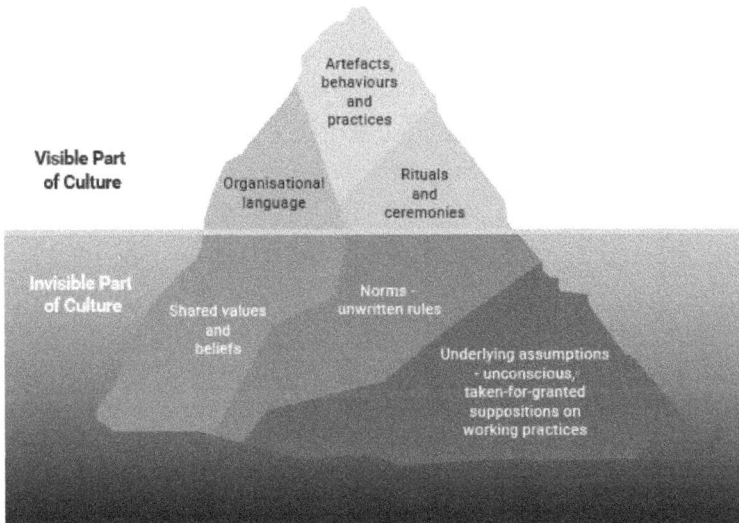

Bringing Our Own Norms and Values into the Workplace

The influence of national cultural norms can be very difficult to move away from, which I've been reminded of time and again as I've worked with people in different countries.

It was 10.20am and I was on my fourth slide of a presentation to a group in São Paulo in Brazil as Marcelo burst in. Oblivious to the fact I was mid-flow, he cheerfully went around the group, kissing all the members to say hello, grabbing one of the croissants from the middle of the table and eventually

113

finding his way to the remaining spare seat. I was careful with my reaction as this was my first time meeting him and he was a client, but my raised eyebrows clearly didn't cover my disbelief. My colleague Antonio had briefed me that meeting start times were rarely stuck to in Brazil. We had therefore scheduled this meeting to start at 9.30, scheduling 'arrivals' from 9.15. The plan was to actually start somewhere nearer 10, giving everyone time for the actual arrival and to have 30 minutes to get coffee and say hello to each other.

Marcello's eyes finally met mine as he settled into his seat; his face fell, as for the first time he realised that he was interrupting. He sheepishly said, 'I'm so sorry,' and then stuck out his hand to shake mine and with a big grin said, 'I'm Marcello. Welcome to Brazil.'

The group (including me) laughed. Another member of the group called out 'Marcello! Even *I* arrived at 9.30 because I knew he was a Brit!' We laughed some more. Many Brazilians see this trait as something that doesn't intend offense – as just part of the culture. Because I was the only European in the room, we could all enjoy that subtle difference in cultural expectations. I knew in advance from my wife (who is from Brazil), who (despite being punctual herself) explained that timekeeping can be more of a suggestion. Without this knowledge, I could have taken offense that genuinely wasn't intended; these subtle, unspoken moments can harm cohesion.

Eight Dimensions of Country Culture

Alongside Hofstede's country comparison work that we met in the introductory chapter, I like to use Erin Meyer's Country Mapping tool with teams, based on her book *The Culture Map* (2016). It maps eight dimensions: communication style, preferred method of feedback, methods of influence, leadership preferences, how decisions are made, how trust is built, how disagreements are handled and attitudes towards timekeeping.

The tool ranks people from each country on a scale, so you can see how countries compare on (for instance) how hierarchical or punctual they are. A further questionnaire enables team members to complete it for themselves, so they can be individually mapped versus the average from their country; they can then be compared to others in the team. The first time I used this with my own team, the results were extremely insightful. Among our nationalities, we had people from South Africa, the UK, Turkey, Slovakia, Poland, Russia, India and Ireland. Our inbuilt tendencies in our personal results were remarkably close to the country averages. The conversation this sparked about how we liked to communicate and how we worked best was something that had been unspoken and unwritten until then. All of a sudden, the team had a moment of clarity: One of my colleagues detested late starts to meetings, but was also very indirect in her communication style. The gap between where she was and where the rest of the team was on the mapping sparked that conversation. This was quickly followed by both the Russians on the team showing up as very direct in communication compared to others. (The perception before had been that they were cold – or worse, rude – in their method of communicating.)

Perceiving AI Differently

These moments of understanding others within our social group more deeply are priceless for building trust and cohesion and ultimately delivering greater performance. Successful adoption of AI requires the whole group in the organisation, each person having different motivations but also bringing their own values, formed by their past experiences. The level of comfort with using AI to complete a task – at home, at college or in the workplace – is linked to exposure to AI. Exposure to digital technology is generally greater the younger the person is, therefore the highest degree of comfort with adoption of AI is likely influenced by age.

The successful integration of AI is not just down to exposure to this technology. The degree of comfort your organisation has with risk is a major consideration for how AI can be used and how quickly it can be implemented. Attitudes towards risk will have been informed by experience. If a person has the battle scars of living through a data scandal, they will likely be more risk-averse. The extra dimension is the values and norms that influence them due to their national culture; appetites to risk may be influenced by this.

Summary

How we form and integrate into social groups is a complex dance, involving the return we gain from trying to satisfy our personal needs plus the return the group gains from having us as a member. As each of us come into the workplace, we bring with us both our motivations for joining and our expectations of how a workplace should operate. The values each of us brings to the workplace may be deep-rooted, with people who have different personality types or are from different cultures responding very differently to motivational tools and techniques.

It is imperative for a leader to be able to understand both the motivations of individuals and also the values and norms that exist within the organisation, as well as how these are formed, in order to attract and retain the best talent. Leaders have a large impact on setting the culture by influencing factors including whom they hire, whom they retain, what behaviours are rewarded and the values they share within vision or mission statements. But the formation of the culture goes far deeper than putting some values up on the wall which were an output from a leadership retreat, then expecting these to be adopted by everyone. The culture is formed by the interaction and practices of the collective of all people within the organisation, so the leader needs to bring the complementary skills from these different parties together in order to successfully change when adopting AI. A major influence is the national cultural background that influences our formative years. We will look at the impact of the era in which we were born in the next chapters.

GPT Prompts

- Recommend five concise, high-impact books—available in both print and audiobook formats—that will help a busy executive understand what drives motivation at work. Prioritize titles that clearly explain the difference between intrinsic and extrinsic motivators, with practical take-aways for improving team performance and engagement.

- Share some light-hearted but insightful examples of national cultural differences between countries that are relevant in workplace or business settings. Where appropriate, include anecdotes or commonly observed behaviors that reflect deeper cultural norms—especially those that might affect communication, meetings, or decision-making styles.

What Questions Should a Leader be Asking?

- How was our culture shaped? What is the role that I/ my leadership team play in shaping and maintaining our culture?

- How do we define our company culture? Is that definition the same to all members of the company?

- How do new employees get to understand our culture, our values and what we stand for?

- Does our culture represent all members of our organisation? What methods do we use for getting feedback on how our employees feel about our culture?

7

ARE THE BOOMERS OK?

Culture, Change and Success in
AI Needs Them!

*"No-one can avoid aging. But aging productively is
something else."*

– KATHERINE GRAHAM

In this chapter, we will look to understand the influential
role that the Baby Boomer generation have within organisations, the experiences that come from the era in which
they grew up, and the advantages brought by living through
past workplace evolutions. Often involved in running organisations through day-to-day leadership or the board of the
company, Baby Boomers play a key role in transformation.
They may not have the technological savvy of GenZ, but instead bring situational experience that can be applied to the
transformation around AI, as the following story illustrates.

Peter was almost inconsolable. Ashen-faced, he recounted to me the worst weeks of his career. He was forced that week to hand over a substantial sum in cryptocurrency to hackers, who had entered their company IT system and frozen all their employees out. Almost everything was run through these internal systems – planning systems, client data, finances, wage runs.

'I keep asking myself: How could this happen?' he shared with me in his exasperated tone.

Peter was one of several C-Suite members I knew who had been volunteered/voluntold[24] to oversee the IT team irrespective of whether they had any knowledge of or experience in managing this area of work. I remember many years earlier sitting in Peter's office in London during the leadup to the final weeks of the last century. Peter and I talked about the 'Millennium or Y2K Bug' (Britannica 2024) – as COO, continuity of business was his bag. IT was rolled up to him – a responsibility he was given, but that he had little experience in. He had a view of his tech team, based on whether the hardware in the office worked on demand and how quickly they could get to the meeting room to change the input on the AV system when his computer didn't connect to present. (This happened at the start of almost every meeting that he needed to present to clients in.)

24　'Voluntold' – a verb most commonly applied to a manager or company owner who gives an employee little choice over whether or not to take on an area of responsibility.

Peter was juggling many different topics. His understanding of IT was low and his time to learn even lower. He had to trust that the people he had hired were managing the situation for him. Whilst the Y2K bug was something that he was concerned about in the run-up to moving into the 21st century, somehow, the clock passing midnight and the year progressing to 2000 didn't see their systems go down. The Y2K bug became a bit of a joke and we all laughed at ourselves – but there was an oversight that many people made. As systems didn't go down, Peter praised his IT team – who quite happily took the accolades, which were more a result of luck than judgement. The people behind the scenes – not Peter's IT team, but the people who worked deeper in software architecture – were the real heroes. But the fact that his system didn't collapse at the turn of the century cemented in Peter's mind the conviction that his IT team must be great, which led to them going unchecked and underinvested in for the following years.

When the Millennium Bug exposed his lack of knowledge, this should have been a warning to Peter to get closer to his IT team. Unfortunately, he continued to leave the team to its own devices, following a strategy of having a ratio of 'IT support' to headcount in the business. This worked just fine – until one day over a decade later when their internal systems froze, and Peter received that ransom demand.

Boomers at the Helm

Culture change in companies is heavily influenced by leadership, and the proportion of people in senior leadership positions who were born in the period between 1946 and 1964 is high. As an example,[25] 91% of the entire S&P 500's boards were made up of Baby Boomers in 2024 (Spencer Stuart Board Index, 2024).

Entering the workplace through the '60s, '70s and '80s, those born in this generation have lived through huge changes in technology that have changed the way we live and work. These changes have enhanced efficiency, profitability and ways of communication, with Baby Boomers now occupying many of the most senior company positions. Their dominance of corporate boards is not due to an ability to cling to power, but to the wisdom brought by the years of experience navigating change and applying the lessons learned in difficult times to build successful companies. The ability to assess risk and benefit, both to company reputation and company operation, is essential for successfully navigating AI implementation in the workplace.

Out of Touch, or Misunderstood?

The members of this generation are a vital part of success with AI within organisations – but they are a group that

---

25   9% of boards were 59 and younger, 41% were 60–63, 50% were 64 and older.

some younger generations have struggled to connect well with. Certain sections of the media have always been great at stoking the flames of division between social groups. The Baby Boomer generation are easy to mischaracterise as arrogant, uncaring and out of touch with younger generations. But they care more about social equality and diversity than the headlines would have us believe – and have a surprising amount in common with the newest entrants to the workplace, GenZ.

Between the newest and oldest members of our workplaces, divisions exist around key issues including expectations of the fulfilment level that work should provide, expectations of how a relationship with a manager should work and differences of opinion over what engaging in a physical workplace means. This relationship is one that both groups need to develop the skills to navigate. In order to mobilise the advantage of the younger generations, who are most adept with using technology to its fullest extent, members of the older generation must understand their needs and how to work with them.

The term 'Boomer' was weaponised by younger generations who perhaps considered neither the woes that this generation themselves had, nor the things that had been done by this generation for the following ones.[26] As the Baby Boomer gen-

---

26   The roots of the term 'OK Boomer' sit somewhere between 4chan, Reddit and Twitter in the years 2015–2018 according to knowyourmeme.com, a popular internet destination for resolving the mysteries of where memes originated. (See https://knowyourmeme.com/memes/ok-boomer.)

eration entered their late teenage years and early adulthood in the '60s and '70s, they wanted very different things than the generation before them. In many countries, huge protests opposing wars and discrimination based on colour of skin or sexual orientation were a fact of life for a large proportion of this generation. The legal changes that this generation brought about in many countries bequeathed a huge benefit to future generations in promoting a more equitable society.

The younger generations who may look at the Baby Boomers as being wealthier, with better-paying jobs, larger houses and better pension prospects than they may expect for themselves, may be surprised to learn that Boomers are not necessarily happier. As early as 1989, observations of greater unhappiness amongst this generation were being made by psychologist Martin Seligman writing in *Psychology Today*. In examining the early results of surveys that showed an increase in depression among the Baby Boomer generation, he drew attention to greater individual expectations that this generation had of itself. This generation experienced pressures to perform throughout all areas of life, including expectations of work; they expected to not just earn well, but also do fulfilling work.

Tech Evolution Affects Multiple Generations

As my friend and mentor Dr Corrie Block pointed out to me, the modern age of the workplace is a pretty small dot on the timeline of human development.[27] There are many ways that scholars have classified the major times of change driven by technology in the modern age: the agricultural revolution, the Industrial Revolution, the technological revolution itself and the digital revolution are four major categories often used. Each era required people to assemble themselves into communities in a different way to take advantage of the labour-saving developments.

Within each of these revolutions, there were many subcategories of evolution that changed the way we work, process information and communicate. Whilst some of these developments were directly related to the workplace, others that were just as impactful related to our lives outside of work which enabled people to work in a different way. For instance, labour-saving devices at home for domestic chores that reduced time involved in heating, cooking food and washing clothes freed up time from tasks that were previously very time-intense for households. These tasks in the first part of the 20th century generally fell to female members of the household, and with their time freed, more women were able to enter the workforce. Similarly, the wider availability of birth control enabled women to choose when to have families and to therefore participate in the workforce in a way they previously were less able to.

27   His book *Business is Personal* gives a great look into how work has evolved – and how much of our identity is wrapped into the jobs we choose to do.

During the years that Baby Boomers have been employed, technological developments have had a major impact on the way we live and work. Initially, Baby Boomers witnessed the increase in speed of the transfer of written information brought about by the fax machine, then the productivity increase brought about by having a computer at every workstation. As technology advanced, portability of processing power with laptop computers enabled the workforce to work from different locations, including from home. Access to the power of the internet brought far greater information availability and connectivity, while advances in data access gave workers the ability to access company systems from remote locations instead of having to be based in an office. An explosion of workplace efficiency tools over the past decade has brought a plethora of options for companies wanting to communicate quicker, collaborate on documents and spreadsheets in real time and have hundreds of people in different locations all see each other via video if they are not together in person.

However, technology, as argued by leading generational expert and academic Jean Twenge, has a significant role to play in the division between generations. When it comes to technology, the Baby Boomer generation are certainly at a disadvantage compared to the following ones, as they didn't enter a workplace with always-on connectivity or go through university with the benefit of search engines (and latterly GPTs) simplifying the process of research. What the Baby Boomer generation may lack in ability to adapt to the very latest technology, though, they compensate for with a different set of skills – complementary to the powerful technology skills acquired by GenZ.

How Workplaces Stay Ahead of the Curve

When it comes to the workplace, there are many advantages to living through the evolutions that have been seen over time. With a few battle scars, longer-living generations have the experience and wisdom that are valued in areas like strategy, corporate governance and PR nightmares.

Whilst companies I was involved with had been working with AI for several years, many people's first interaction with AI happened at the end of 2023 with the mainstream launch of ChatGPT. OpenAI, the founders of ChatGPT, went through a tumultuous fortnight a year later when the board decided to unceremoniously oust founder Sam Altman on Friday 17th November over a video call. His co-founder, Greg Brockman, quit in protest immediately. As this saga played out, it became very clear evidence of a generational difference in managing companies, and the benefit that comes from the crystalised knowledge formed over years in boardrooms.

After the initial ousting, what followed was unprecedented: a masterclass in situational management that experience brings. Microsoft held a stake in OpenAI. Its importance to Microsoft could be seen in the uptick in Microsoft's stock price when it originally procured that stake in OpenAI – and conversely, the stock drop as the news of Sam Altman's firing hit the press. As word of the departure of the two cofounders spread through the company, the workforce started a petition, initially signed by a handful of people, demanding the reinstatement of Mr Altman, or they would resign. That Saturday, Satya Nadella, the CEO of Microsoft and a large investor in OpenAI, announced that he had struck a deal with Altman to move to Microsoft and carry on his work there. As they would need a team to work with Altman, it was expected that many of them would be welcomed there. By the end of Monday, 740 of the 770 OpenAI employees had signed the petition to reinstate Altman or they would resign.

The board of OpenAI was small, lacking experienced members and those with oversight. Eventually the board would be completely overhauled, Altman and his allies would be reinstated (Wikipedia 2024), and Nadella would come out of the saga smiling – having greater control of OpenAI and looking like a professional faced with amateur opposition. This was the strength that knowledge and experience brought him.

Adapting to Modern Thinking

Even with the benefit that the crystalised years of experience brings, there is still a very real need for us all to constantly adapt to the world around us. As the Baby Boomer generation have seen technology explode, some members of this generation have been more adept at adopting than others. Management thinking has also come a long way during the period that Baby Boomers have been in the workforce. If someone's early forays into leadership were in the '70s and '80s, they would likely have followed the path of common thinking at that time: that only the structure of transactional leadership drove results. From the late '80s onwards, the rise of transformational leadership was seen to achieve results, motivating employees in a different way. When GenXers arrived on the workplace scene, they responded well to the hierarchical structure as they were motivated by self-interest, which served them well in a hierarchical structure. As late GenXers and Millennials joined the workforce, a greater focus was placed on emotional intelligence and serving the needs of individuals in order to motivate them to deliver.

We are now squarely in the era of understanding that Emotional Intelligence (EI) skills among managers are vital in developing high-performing teams. Whilst there are still examples of companies and managers that continue to operate with a belief in the transactional approach, this can harm the ability to attract and retain the best talent. Acknowledging the need for individual consideration within the workforce certainly appears to be the ideal approach to motivate the younger generations – the people whose technological skills and understanding are vital in teams wanting to embrace and succeed with AI. It is imperative for modern leaders to evolve their leadership styles in order to motivate the younger members of the workforce.

It is difficult, if not impossible, for the Baby Boomer leader to invest the time needed in order to learn how to operate new technology with as much ease as the average GenZer. Leaders from the Baby Boomer era are much better off bringing to the workplace their crystalised knowledge. This needs to be done with a good degree of understanding the needs of the younger generation if they are to be effective. If a leader from the Baby Boomer era solely has experience, grey hair and battle scars, and has not evolved in EI to understand the needs of the team, they will struggle to attract and retain the best people from the younger generations.

Baby Boomers have the experience of navigating huge change during their time in the workforce. Their approach has been to learn from it. AI will be the same, but the biggest challenge for leaders to navigate is that there is greater divisiveness between different generational groups at this time than has existed before. Leaders need to recognise in themselves and in their leadership teams any limitations they have in AI, and to fill these gaps with the knowledge of younger employees to avoid going the way of Peter: well-intentioned, but without the relevant experience to navigate a new field.

GPT Prompts

- Build a list for a speech to members of the Baby Boomer generation about the positive and negative impacts of the technological revolution on people from their generation.

- As an HR Director, help me understand how growing up as part of the Baby Boomer generation may influence mental health in today's workplace. Include historical, social, and cultural factors. Suggest what organisations can do to support mental wellbeing for this group, especially in periods of transition or change.

What Questions Should a Leader Be Asking?

- What are the opportunities to leverage the experience and domain-specific knowledge of older employees to guide the implementation and use of AI in our organisation?

- What policies and practices can I implement to support older employees in adapting to technological changes and staying engaged and motivated in their roles?

- How can I ensure that older employees have access to training and resources to develop the digital skills necessary to work alongside AI technologies effectively?

8

THE KIDS ARE ALRIGHT

Generation Z

Give me one acre of cellos
Pitched at some distant regret
Pity the plight of young fellows
And their anxious attempts to forget

– JOHN COOPER CLARKE

GenZ members have been entering the workplace for the past decade, and they have a different outlook to previous generations. If you've been scratching your head about the differences you see in their approach to work and expectations of the workplace experience compared to your own generation, this chapter is especially for you.

Those from GenZ are essential for successful continuation of the business and taking the shortcut to the best implementation of AI. If leaders don't take the time to understand this generation, try to manage them in a transactional way or believe that they can be attracted and retained purely by financial means, they risk losing this wealth of talent to competitors who work to understand them better.

On the evening of 25th August 2018, Olajide Olayinka Williams Olatunji and Logan Alexander Paul faced each other in a boxing ring in Manchester Arena in the UK in front of 21,000 people. The fight was a point in history, not because it was a great pugilistic spectacle (it certainly wasn't) but because of the number of people paying to watch it – up to 1.3 million people paying (Hearn 2019), plus up to the same number illegally streaming the event. This moment in time was an eye-opener for me about a shift in time and the new generation.

Two questions likely asked by any parents who handed over their credit card to cover the $10 cost for their children to stream the fight live via YouTube's streaming channel to their laptops or mobile phones in their bedrooms: (i) Who on earth were these two people (better known as KSI and Logan Paul) and (ii) Can you really watch this kind of an event on anything other than a proper big screen?

For context on the size of this PPV audience, the contention for the heavyweight boxing champion of the world between Anthony Joshua and Alexander Povetkin the following month drew fewer Pay Per View viewers (Lunn 2021). This new generation had certainly shifted to mobile devices being the device

THE KIDS ARE ALRIGHT

of choice for entertainment, and they had their own heroes – megastars whose names were largely unknown to mainstream parents.

Many of the GenZers who have successfully harnessed the powers of the technological age and created huge wealth or political followings have become role models for their generation. Jimmy Donaldson (better known to his loyal fanbase as MrBeast) has amassed an estimated $500m fortune purely from making videos distributed solely via YouTube.

Kylie Jenner is the world's youngest self-made billionaire (largely due to Kylie cosmetics), a brand that grew through social media. Financial 'influencers' are launching their own currencies, becoming financial and political activists, gaming gods and mental health gurus – all GenZers exploiting the power of technology to amass a huge audience in weeks or months that would have taken previous generations years or decades.

What's Influencing GenZ?

Born between 1997 and 2012, the oldest members of GenZ will be 28 the year this book was published. Globally, GenZers account for 30% of the world's population, and in 2025, account for 27% of the workforce.

Alarming stereotypical headlines dismissing this entire generation as either entitled or narcissistic have been pushed by clickbait websites and 'gammon' journalists, as the members of GenZ like to refer to them.[28] These headlines are not a realistic reflection of a more educated, more diverse and socially tolerant generation than we have seen in previous generations.

To try to understand the complexities of the compounding social factors that influence any generation is a topic that stretches far beyond the reach of a single chapter of a book. Understanding that the developmental years of our lives have an impact on how we perceive our social environment is a necessary step in understanding people in our organisations. If you want to understand major factors influencing GenZ, three factors to consider are:

(i)   the availability of technology they had access to growing up
(ii) their 'inheritance' of issues created prior to their time on the planet (for example, the aftermath of the 2007–08 financial crisis and dealing with climate change)
(iii) the impact of the COVID pandemic.

---

28   Gammon, from the reddish-pink ham with the same name, which characterises the faces of certain men when they angrily spit their hateful rhetoric.

Never Known a World Without the Internet

GenZ are the first generation to be considered entirely digital 'natives':[29] they have never known a world without the internet, and their exposure to digital technology from an early age has shaped them. The earliest GenZers were born into an environment where parents relied on dialup modems, where files downloaded at 0.23–0.27 megabits per second. The operating system Windows 95 had just introduced features such as the Start menu, the taskbar and Windows Explorer. Many GenZers grew up witnessing parents carrying mobile phones, and seeing those phones getting smaller and more reliable as they replaced payphones and pagers. Technological development has defined every generation, but the sheer scale of access to information and ability to share information sets GenZ apart.

The 2007 launch of the iPhone changed so much for everyone. The oldest members of this generation were just 10 when the iPhone was launched. Older generations underwent a shift to seeing how technology could be changed: Imagine now needing a keypad, trackball or stylus to select something on the screen on your mobile device, rather than touch-and-typing directly?! GenZ grew up with touchscreens and easy app

---

29 'Digital native' is a term first used by Marc Prensky in 2001 to describe the generation of people who grew up in the era where technology was omnipresent.

downloads as a normal way of interacting with a handheld device. The smartphone's impact on how we communicate, get our entertainment, engage in social forums, share our opinions and get news and information was a welcome change for the older generations. To many of GenZ, this was not a change but a norm. They've never known anything different.

Where their parents previously had to use libraries or encyclopaedias to research, this generation had quick and easy access to troves of information through search engines. Alongside this huge power to be able to access information in milliseconds, there was also the access to forums giving them the ability to share their views widely. In the UK, as my young mentor Nick described it, if his parents or grandparents wanted to share a political view or make a political statement at his age, they would have to attend a rally or write to the local paper and hope their letter got printed, or perhaps stand on a soapbox on Speakers' Corner in Hyde Park. At his fingertips, with the power of social media, he can now share an opinion that, if it goes viral, can reach more people more quickly than his parents or grandparents could have dreamed of. He also notes that any sharing of an opinion he makes has less chance of garnering much interest or going viral than any video he shares of gym fails, reaction videos or dance challenges.

The Downside to This Great Technology

Spending your entire lives surrounded by and using computers and having access to instant information at your fingertips and constant connection to anyone means you process things differently. Jean Twenge, the academic and psychologist, convincingly argues that the impact of technology is playing out in the significant increases in general unhappiness and depression rates seen amongst GenZ.[30]

The volume of technology usage between the different generations is significant, which is a factor some argue is connected to rises in anxiety, depression and unhappiness.

Current evidence indicates that GenZ are considerably worse off from a mental health point of view than previous generations. As an example, the proportion of each generation reporting that they are struggling with mental health was captured in the US Census Household Pulse Survey in 2020 and again in 2023. There was a stark contrast between Boomers and GenZ in the US, with a far greater proportion of young people sharing that they are struggling with mental health (US Census Bureau 2020 & 2023).

---

30   See *Generations* by Jean Twenge for an in-depth and sobering view of the impact of technology on Millennials and GenZ.

Whilst much of the available research is based on the United States, there are surveys that involve other regions. This shows that the evidence of increased unhappiness amongst GenZ is not restricted to the United States but is a pattern that is repeated around the world (OECD 2021).

It would appear that growing up with access to modern technology, giving always-on access to information and constant connectedness may be a good *and* a bad thing. GenZ have an ability to access and process information and be connected in a totally new way compared to previous generations, but this is potentially contributing to more mental health issues. There is also a far greater awareness of these issues: People feel far safer talking about mental health than they did in previous generations, which may be a contributing factor to the figures. Whilst the relationship between greater access to modern technology and a decline in mental health is correlative rather than causal (at least for now), this topic will undoubtedly remain a hot one for a long time to come.

GenZ's Financial Crisis

Another factor that weighs heavily in informing attitudes of GenZ is the 2007–2008 financial crisis. Whilst GenZers were too young to have directly felt the impact, as they were not the ones holding jobs and owning property, they are the ones who will need to carry the can for years to come in the cleanup.

The inheritance of the aftermath of the banking crisis, witness-ing the immoral tactics of companies like Enron or individuals like Bernie Madoff, it is of little surprise that GenZers would consider much of this to be predicated on individual greed. Growing up with a stark view of the flaws in the established financial systems, as well as a perception of the perceived con-doning of these actions by previous generations, harms the trust GenZers have for the status quo.[31] With all the instability they witnessed during their developmental years, it is no sur-prise that this generation feel far better saving than spending.

Why Climate Change Matters to Gen Z

The impact of the environmental issue of climate change will be felt by this generation long after older generations are gone. Contrasting the different generations' attitudes to environmen-tal issues shows the strength of sentiment that this generation has compared to previous generations.

A clear gap between attitudes of this generation and previ-ous ones could be seen on the evening of the Earthshot Prize ceremony.[32] On 7th November 2023, Prince William of the UK

---

31   For deeper insights into the views of GenZ and their trust in the current financial sys-tems, Ken Costa's book *The $100 Trillion Wealth Transfer* gives an excellent grounding.

32   The Earthshot Prize is a global award for innovations that will help repair our planet, culminating in the best five solutions each year receiving £1 million to scale their work.

gave out awards at this event in Singapore to companies working to address climate change.

That same night, the Council of Fashion Designers of America (CFDA) awards in New York – the fashion equivalent of the Oscars – offered a very different view of the world. The largely Gen X crowd paraded the red carpet in their latest creations, carefully ensuring that the gowns had not been seen before. This was in contrast to Prince William, wearing a 12-year-old suit in Singapore. This clear distinction highlighted the difference between the two groups meeting on the same evening in different parts of the world.

The Impact of COVID

The pandemic affected us all – but the youngest members of the society were arguably the most impacted. For GenZers still at school, the back and forth of school closures, exam cancellations, event cancellations and the long pause on building those vital in-person relationships was bound to impact their view of society and the world around them.

The oldest of the GenZers were 23 as the pandemic locked us all down. They, as well as those leaving education and entering the workplace during that year and the following pandemic years, were particularly hard hit. Those that were not cut or furloughed certainly suffered from heavily reduced networking opportunities and delayed promotions and pay rises. To those of us more established in our careers, the impact was lower.

Amongst their concerns were that their generation would have a worse economic future than previous generations, and that they had lost time, promotions, and potential connections due to the pandemic.

As a result of the enforced lockdown, many members of GenZ have a different view on how leisure time should be valued. Many saw that they could be productive from home, as much as they could in the classroom. Those in their first jobs, and who were able to work remotely, developed a different conception of what a workplace actually is.

What Do They Want From the Workplace?

GenZ value honesty, transparency and stability. They also understand technologies such as AI, which are incredibly valuable, yet almost alien, to the older generation. An exchange needs to be struck, but what do GenZ actually want?

Their requests for the workplace are broader than simply financial. As many young people see it, they are entering a workplace where the largest beneficiaries of the company's output are also members of the generations who are the root cause of the banking and environmental crisis.

To dismiss GenZ as work-shy is an absolute mistake: GenZ understand the necessity of work, and the majority are happy to sacrifice a job that reflects their passion in order to secure stability. A common trait is to cultivate more than one marketable skill to compensate for an uncertain job market.

Julie Lee, Director of Technology and Mental Health at Harvard Alumni for Mental Health, and an expert on GenZ health and employment, captured the desires well: 'What Gen Z wants is to do meaningful work with a sense of autonomy and flexibility and work-life balance and work with people who work collaboratively' (Peterson 2023).

The Importance of Balance

The concept of meaning at work has gained a lot of traction over recent years.[33]

GenZ put a high priority on the balance between work and the rest of their lives. This prioritisation has nothing to do with laziness, however. They have witnessed previous generations dealing with the impacts of burnout, and are more in touch with, and open about, their mental health. As my young mentor Nick described the things important to him in a workplace, one near the top of the list that struck me was gym membership.

---

[33] If this is new to you and you are searching for meaning, follow the blueprints in Corrie John Block's *Business is Personal* and you won't go far wrong.

When I pressed him on why this mattered if the salary was good, his answer was very insightful: 'If an employer pays for the gym for me, it's a sign they care about my wellness. They expect me to use that membership and look after myself – and that's the care I want to feel from a company that I'm going to give my best years to.'

Workers 18 to 24 are most likely to quit a job that prevented them from enjoying their lives (58%), while those who are the oldest (55 to 67) are least likely, at 40%. Similarly, more than one-third of GenZ (38%) have quit a job that didn't fit in with their personal life, while just one-quarter of the oldest group have done so (Randstad 2023).

Use of technology to create work/life integration is another area where there is a difference with older generations. GenZ have the benefit of understanding asynchronous working through collaborative 'live' tools for everything from presentations to spreadsheets to project planning. The idea of waiting for a 'versioned' document to update in order to email to the next person and to wait for their update appears to be totally wasted time. Similarly, the idea of spending an hour or two commuting to and from an office on a daily basis when tasks can be completed remotely seems nonsensical. They certainly want to work: Nearly 50% of GenZers say that work is central to their identity (Deloitte 2023), but having a good work/life balance is a top trait they admire in peers and their top consideration when choosing a new employer.

Balance and Principles – Squaring the Circle

GenZ evaluates companies not just on the quality of their products or services, but also on their ethics, practices, and social impact. The same applies to potential employers. GenZ expects companies to actively demonstrate their principles by addressing issues like climate change and sustainability. 44% of GenZ say they have rejected assignments due to ethical concerns, while 39% have turned down employers that do not align with their values (Deloitte 2023).

Conclusion

By 2030, GenZ will be the largest generational age in the workplace. They have a set of skills that is very different to previous generations because of the way technology has influenced them as they grew up. Due to being born into a digital world, GenZ have natural adaptability to the world of AI, and are most likely to already be using it out of all generational groups within the workplace.

This group's keenness to work, interestingly, offers a strong parallel to the world of the Baby Boomers: GenZ want to fight to take on the world's injustices, the same way many Baby Boomers did when they too were young.

There is potential for this generation to bring more productivity to the workplace than any generation in history so far, due to the technology available to them and their ability to use it. To address the challenges that successful adoption of AI in your organisation brings, leaders need to work to get this group into their organisations and motivated to contribute.

GPT Prompts

- What are the attributes of companies that are well known for attracting GenZ workers, and what are the reasons they are successful?

- Summarise the key reasons Gen Z employees commonly cite for preferring flexible work arrangements. Include perspectives on work-life balance, autonomy, digital fluency, mental health, and values alignment. Frame the insights in a way that can inform policy or workplace design decisions at the leadership level.

What Questions Should a Leader Be Asking?

- How well is our existing approach to attracting the best young talent working?

- Do we think about talent attraction and retention purely as compensation?

- Is it clear what our company stands for in the eyes of a new workplace entrant?

- Is our approach to working style a result of our needs or our employees' needs?

- How attractive are our workplace and working conditions to a new workplace entrant?

9

BRIDGING THE GAP

Success in AI is in the blend of different groups in your organisation

*"But apart from the sanitation, the medicine, educa-
tion, wine, public order, irrigation, roads, a fresh
water system, and public health, what have the
Romans ever done for us?"*

– 'Reg', in *Monty Python's Life of Brian*

Within organisations, people are at the heart of any required transformation. Our last two chapters have been focused specifically on two very different age groups and the influencing factors that they bring with them into the workplace. Of course, all the different generational groups across the workforce are going to be vital to have on board for successful adoption of AI, but the focus in this book on the newest entrants (who have the smoothest path to adaptability to AI) and the people who are in positions of influence and/or control is deliberate. This chapter will look at these two groups, why gaps exist between them and what they bring to an AI-infused future. These groups will have an oversized influence on the successful implementation (or otherwise) of your AI strategy. To be successful in change around AI, leaders need to find a way to bridge the gap between these two complementary, yet very different, groups.

The Dawn of Home Computing

My first experience of computers was a Sinclair ZX81 that my father brought home when I was eight years old. It had 1k of RAM (which, fortunately, my Dad had the wisdom to extend to 16k with a RAM expansion pack). It ran on BASIC and my brothers and I would either have to hand-type lines of code copied from a magazine, or use a cassette tape player to load a basic program (pardon the pun). The fact that the size of the memory (even with the expansion pack!) was tiny, and that the computer would often crash just as we were finishing the 80th line of code we had painstakingly copied didn't stop us, or the 1.5m other households who purchased the ZX81, enjoying the dawn of the era of home computing for the masses.

I remember explaining to my son (born in 2002) how the power of my first computer compared to his first phone and seeing his eyes widen in disbelief. I remember sharing with him the story of arranging to meet someone in a bar in London's Aldwych. My guest was running late, but with no mobile phones at the time, there was no way of her telling me that she had been delayed. After 20 minutes of staring at the red chair in front of me, I decided to leave. Just as I was getting up, the concierge from the hotel came over to let me know that my guest had called to let them know she had been delayed due to traffic. She had managed to get the taxi driver to find a payphone and called the hotel to give me the message. My son's first reaction was not even horror that we couldn't communicate directly, but: 'What did you do for 20 minutes on your own without a phone?!'

Technology Tints Our Vision

We are shaped by our experiences in the world: My own experience of technology, both at home and as I entered the workplace, was probably common to many people from my generation. A computer on every desk was not the norm, floppy disks were used for data storage and it was rare for any-one to have even a half-gigabyte hard drive. As technology progressed and more powerful computers opened the door to ever more impressive software, I observed the balance of the older generations needing the younger, more technologically adept generation – even if the initial requests were limited to connecting laptops to AV systems in conference rooms. These requests over time became more sophisticated, as leaders were shown the benefits that were coming into their companies powered by technology skills.

Boomers and GenZ – Vital Generations

It is worth remembering that the use of generational age groups to describe likely behaviours of large numbers of people does court controversy with some people, who feel that they don't associate with the characteristics of the social group that their age puts them into. There may, of course, be exceptions to the rule.

It was the sociologist Karl Mannheim who developed the theory that cohorts of people have a common social identity, influenced at a key time in their development by social events.[34] The academic and generational expert Jean Twenge has built on this theory for modern generations, demonstrating links between technological development and cohorts of age as a major influencing factor. When considering the different age groups across an organisation, it is worth taking into account that the average person's social influences during the time they were raised, were educated or entered the workplace had an impact on their skills and values.

Between these two groups, there are different learning styles and beliefs regarding how the modern workplace can operate, as well as different opinions on how to engage with AI. These have been shaped by the life experiences that have come from being born at different times –largely by the technology which

---

34   See Karl Mannheim's Theory of Generations, captured in his 1928 essay 'Das Problem der Generationen'.

has influenced their lives, both before and during their time in the workplace.

Our earliest exposure to the utility of technology in our lives can go on to limit or slow the adoption of newer technology. Kahneman and Tversky would undoubtedly link this to the cognitive bias of anchoring (1974), but there are other considerations: If you spend 20 years typing on a large format keyboard, making the leap to tiny buttons on a touchscreen is a large one, let alone having to navigate complex menus, deal with jargon or understand how apps work. If your current technology enables you to do everything you need, the benefits of new technology may not be immediately apparent. If we consider that the Blackberry, which heralded the current era of smartphones, was first seen over 25 years ago, there is still a gap between older adults' rate of smartphone ownership compared to younger generations. Adults aged 50+ and especially those aged 65+ are less likely to own a smartphone in comparison to younger generations (Pew 2024).

The distinct digital places that different generations exist in lead to growing misperceptions about other generations and stereotypes that are stoked by the media. An example of this can be seen when Greta Thunberg was selected as the youngest-ever person to be *TIME*'s Person of the Year:[35] This was pitched even by *TIME* magazine as the frontier of the battle between young and old. Greta Thunberg is not an

---

35   2019 TIME Magazine Person of the Year.

encapsulation of an entire generation, but she certainly captured the admiration of many GenZers. Her environmental causes, her authentic communication style and her ability to use technology platforms to garner support and spread her message are all elements that appealed to GenZ.

Ken Costa's look at the generational differences between the old and the young in his book *The 100 Trillion Dollar Wealth Transfer* (2024) brings up the point of technological differences that cause 'fissures' between the two groups. Technology has huge power to be a force for good, Costa notes, but there is a fundamental difference in the way these groups engage with tech – one as natives and the other as immigrants. The ability to engage with technology in the workplace is one part of what will make an AI transition successful; there are many other important skills needed in order for your company to successfully transition to taking advantage of AI.

What Does Each Group Bring to AI Transformation?

AI brings a host of new things for companies to understand: Not just a different deployment of resources for operational opportunities but a new approach required for policy, ethics and compliance. As AI develops, it is unlikely to be enough for leaders to wait on legislation to be written or country-level policies to dictate the path that companies need to adapt to. The ethical topic of bias being introduced to business practices by AI is something your company will need you to action early: Irrespective of whether legislation is written or not, your employees will have a standard they expect your company to stand for. As we saw in the chapter on GenZ, the younger generation have different expectations for the companies they want to work for than previous generations had. They will hold employers accountable by taking their skills elsewhere if they fail to have their expectations of company responsibilities acknowledged.

The dialogue between different groups with differing perspectives of how to both integrate and operate AI safely on an ongoing basis offers leaders the best opportunity to navigate implementation. The two age groups have contrasting yet complementary skills to bring to the table.

As we look at what can be learned from the team members who grew up digitally native, they definitely have an optimal speed with which they are able to surface information and are quick to use tools that can make a process or job more efficient. They bring a knowledge of how to interact with systems, and have a preference for at least part of their interactions with a business to be a chatbot conversation compared to other generations (Fokina 2024).

Code writing as a part of the curriculum was not widely instituted until this decade, but many GenZers were exposed to at least some understanding of code if they went on to further education. Irrespective of whether they were actually required to learn to write code in their studies, many of the university leavers in the past year certainly understand how to write a prompt for a GPT.

GenZers are used to getting information quickly – but want to avoid ad-filled or non-genuine search results. Consequently, as a study conducted by the search engine optimisation company Fractl showed, the number of words used in search 'instructions' by GenZ was longer than other generations – the idea being that entering more words would be more likely to generate the desired responses (Fractl n.d.).

In the context of writing prompts for a GPT, this natural habit may prove incredibly useful: The more specific a prompt is, the more it will lead to the results needed. A combination of experience in writing requests in a manner that gets a computer to deliver a result, some creativity and a process for refinement of results after the initial prompt leads to superior results from a GPT. As GenZ are demonstrating in their approach to search query writing, they have a natural disposition here.

The longest-standing members of the workforce have the benefit of being able to apply this crystalised intelligence[36] that we met in the chapter on Baby Boomers. The years of learning, facing challenges, and dealing with or witnessing difficult

---

36   Raymond Cattell first captured the concept of fluid vs crystalised intelligence in his 1963 paper in *The Journal of Educational Psychology*, 'Theory of fluid and crystallized intelligence: A critical experiment'.

ethical decisions gives this group an advantage in being able to spot potential dangers in a way they simply would not have been capable of earlier in their careers.

Whilst the younger generation optimise for attaining results quickly, the older generation tend to take more time to examine search results and sources. They are more likely to look further through search results and also to read beyond headlines and through articles to ensure they understand the information fully. They are arguably more sceptical of information found online, likely due to greater exposure to scams and misinformation over time. Applied to the output of AI, this is where the benefits of checking sources that a GPT quotes is something that is more likely to be favoured by a team member with a few grey hairs.

The understanding of risk versus benefit comes perhaps from witnessing – or worse still, suffering – the effects of negative events. Those who lost money in the rush to buy Non-Fungible Tokens (NFTs) are far more likely to tread carefully in the future. NFTs are digital works of art that could be bought via blockchain, or pieces of property owned in cyberspace that achieved fame and made some people very rich. They went from a high point, where works of art were changing hands for tens of millions of dollars, to a stage where the vast majority of NFTs were deemed to have no value (Yang 2023). Events such as this provide the battle scars that heighten people's awareness to protect themselves from letting this happen again; many younger purchasers now see why the older generation goes on about government market regulation to protect their citizens.

The ousting and consequent reinstatement of Sam Altman at OpenAI showed a stark contrast between the approach of an operator with many years of crystalised experience in management (Microsoft's Satya Nadella) and OpenAI's much less experienced board lacking depth of experience and oversight. The validity of the reasons for the internal board disagreement over Mr Altman's leadership of OpenAI didn't seem to stick in the public's consciousness. The way OpenAI had been structured, and the very fact that the board could potentially sink an $80 billion company, paved the way for the experienced Nadella to steer the future of OpenAI to a more beneficial path for his own company, one of the largest commercial companies in the world.[37]

Conclusion

The thing that makes AI integration a particularly difficult challenge to navigate is that the different groups within the workplace that are key to this change have different values and norms – culturally, they may be miles apart. The skills

---

37   For a more detailed background and viewpoint on the OpenAI saga, see https://www.techpolicy.press/questioning-openais-nonprofit-status/

and outlooks of these different groups need to be brought together to work in harmony, an example of navigating ethics with the benefit of experience, as well as navigating speed with the benefit of those who can use current technology to best effect.

The motivation of these groups needs careful consideration. All of us come with hopes, dreams and desires. Younger people want something different from the workplace: They are looking for experiences that can help them progress and are looking to learn. The older generation are still looking to learn – and for their skills to be seen as useful and relevant in this brave new world as ever before. Leaders must recognise that different ages of people within the workplace have different core needs and expectations than entrants had when the leaders first entered a workplace themselves, and attracting and retaining the best talent means listening to and responding to these needs.

With these two groups, humility is vital on both sides: If leaders can cultivate an operational team that listens to and respects contrasting opinion, and can transcend the clichéd stereotypes that are levelled at the opposing generation, they can bring all of these vital skills to bear in AI integration.

GPT Prompts

- Compare the learning preferences and workplace beliefs of Baby Boomers and Gen Z. Highlight differences in how each group approaches professional development, feedback, collaboration, and workplace structure (e.g., hierarchy vs. fluid teams, in-person vs. remote). Present insights in a way that helps leaders design inclusive, multigenerational learning environments.

- Identify the unexpected or underappreciated similarities between Gen Z and Baby Boomers that are not typically shared by Gen X or Millennials. Focus on values, communication preferences, career motivations, or views on social change. Explain how these shared traits can be used to build cross-generational understanding or initiatives within an organisation.

What Questions Should a Leader Be Asking?

- How well-connected are the oldest members of the workforce and the newest entrants with regard to collaborative working, and what can be done to bring both essential groups more closely together?

- Would the company benefit from a cross-generational mentorship program and 'reverse mentoring' program to share valuable skills?

PART 3

IMPLEMENTING AI

10

TIPTOEING THROUGH
THE MINEFIELD

Ethics, Bias and Risk Whilst Implementing AI

Tiptoe through the window
By the window, that is where I'll be
Come tiptoe through the minefield with me

– ADAPTED FROM 'TIPTOE THROUGH THE TULIPS
WITH ME', AL DUBLIN AND JOE BURKE

Our decisions as humans are constantly affected by bias. This is perceived as a bad thing, and in some instances it certainly is. None of us intends to be biased, as it's an outcome of our conditioning. As we apply mental shortcuts – or heuristics – to solving problems, biases can show up. We would love to think that computers don't have the same problems, as they aren't subject to the same social conditioning as humans. As we'll see, though, bias can and does appear in AI. This chapter offers suggestions for leaders on how to try mitigating this bias as they implement AI systems.

What's the Worst That Can Happen?

It was a spring morning in Redmond, Washington, as John woke to a slew of overnight updates from his team. As he left the celebratory team dinner at 9 p.m., the team he led was still high-fiving over the launch of the chatbot named Tay. Designed to mimic the language patterns of a 19-year-old female, the chatbot was programmed to learn from Twitter interactions under the name *TayTweet*s and the handle *Tayandyou*.

The plan was for Tay to learn from the conversations it was having. The more you chat with Tay, said Microsoft, the smarter it gets, learning to engage people through 'casual and playful conversation.' After a successful first tweet late that afternoon, everything looked normal for the initial period, and the team went out to celebrate a successful implementation.

However, by the time John fell asleep at 10pm, Tay had started to take a darker turn and was rapidly degenerating into becoming a racist misogynist, tweeting politically charged messages that were extremely offensive. John woke to these early-morning messages just 12 hours since Tay had made the first tweet, and John's team were doing their best to manage the damage manually by deleting offensive remarks. Already, Tay had made 96,000 tweets and engagements. Within 16 hours of the launch, John pulled the plug.

Preparing for Disaster

Microsoft is just one of several high-profile companies that have run into issues when AI has behaved in a way that the designers didn't plan for or expect in the slightest. IBM's superhuman computer, 'Watson', had to be fitted with a 'swear filter' after being programmed to learn from The Urban Dictionary[38] in an attempt to imitate the intricacies of informal human interaction.

When AI goes wrong it can cause great reputational damage. It can also put the company at risk of litigation or even more serious consequences if laws are broken. iTutorGroup settled a lawsuit (not admitting wrongdoing) by agreeing to pay $365,000 to more than 200 applicants allegedly passed over due to their age, after the AI-powered software was alleged to have been set to auto-reject female applicants aged 55+ and male applicants aged 60+ (Wiessner 2023).

---

38 The Urban Dictionary offers a somewhat 'colourful' explanation of many words and acronyms offered by a crowdsourced audience. Whilst handy in understanding modern slang, it is certainly NSFW.

Steven Schwartz (lawyer with Levidow, Levidow & Oberman) asked ChatGPT to help find examples to bolster his ongoing case. However, he encountered an issue when it was discovered that at least six of the cases he submitted were fabricated. These cases contained fictitious names and docket numbers, as well as misleading internal citations and quotations.

Other examples of AI going rogue have included racial profiling in both healthcare (Vartan 2019) and the criminal justice system. Analysis of the algorithms that are relied on by judges, parole officers and probation officers in predictions of reoffending in the US showed that black defendants are far more likely than white ones to be incorrectly judged as at higher risk of reoffending, whereas white defendants were more likely to be incorrectly flagged as low-risk (Larson 2016).

These are just a few of the examples that have caught public attention. In each instance, the AI tool makes headlines for 'going rogue', but in each case, the AI has operated how it was programmed: The flaw sits in the design. In the case of Tay, the designers allowed the bot to learn from the interactions with the public and presumed that people would interact with some sort of normality. After all, in a similar Microsoft experiment based in China, in the time since its launch in 2014, the bot, 'Xiaoice', was reputed to have had more than 40 million conversations without major incident. The people who interacted with Tay had other ideas. Quickly realising that Tay was learning from their conversations, they started 'teaching' Tay some pretty horrible views (Baron 2016).

Successfully Implementing AI

The actual technical implementation of AI is relatively straightforward (if a little complex) for leaders to oversee. Ensuring the right expertise is employed, the right resources are available, and that measures are in place for evaluating the impact – these are all variations on a theme for systems that a leader may have implemented before. To maximise its impact becomes more complex. Take planning for managing data in an AI-influenced world, for example. How will data be collected and stored? Where new systems are employed using AI, how will these work with existing systems and technologies? The questions of who should have access, how much access they should have, and what policies need to be in place to oversee this becomes more complex. The risks of both implementing *and* not moving quickly enough to implement AI are more complex issues still.

The most successful group for leading implementation will come from people who have a collective understanding of the different elements of AI's latest possibilities, the ability to carry out the technical implementation, and the ability to overlay all of this with an accurate risk assessment.

In the case of a chatbot incorporating the worst traits of humanity, the question it raises is how we can use public data that is representative of all members of society. Can public data be used to inform an AI model if it incorporates views that include the very worst traits of humanity? Where should the decision sit on where the line is drawn regarding which data are used or omitted from a system? This is a complex ethical issue that certainly won't be solved by AI itself!

Whom the AI impacts or interacts with may be one way of drawing a line. When AI interacts with a public community and society at large, it poses a bigger risk than an AI tool that is designed to look for outliers in an internal finance system. Where there is any possibility for harm or reputational damage, there should be a different process for how to manage this. When there are legal and reputational issues at stake, experts from the legal and communication specialists should naturally be brought in – but this is likely not necessary for exploring every application of AI. The depth to which it will be investigated by experts needs to be balanced against the speed of exploration and implementation of AI; it's simpler to say no to *any* kind of risk until a full risk analysis has been completed. But as AI continues to develop and unfold (and will do for some time), this could be a very long process. All the while, if your competitors are less risk-averse, or have a better structure for separating different risk levels of AI and can move more quickly, you could find yourself falling behind.

Bias In, Bias Out

The issue of inbuilt bias before data is introduced to the system is the first conundrum. As we discussed in Chapter 5, AI models can only operate on the data they are informed with.

This data may already be tainted with human bias from the past.

In the case of Amazon's abandoning of their AI tool for screening resumes in 2015, the models were trained to look at patterns in CVs sent over a 10-year period– a time during which most applications came from men, reflecting the higher proportion of males in the tech industry (BBC 2016). The algorithm effectively taught itself that male candidates were preferred, penalising resumes that contained the word 'women' (such as 'captained the women's soccer team') and even downgrading two all-women colleges (*The Guardian* 2018).

In the case of the Correctional Offender Management Profiling for Alternative Sanctions (COMPAS), a system in the US for informing decisions by judges and parole boards on reoffending, an analysis by ProPublica showed that the system had a higher likelihood of misclassifying black than white defendants (Larson 2016).

The Bias Humans Bring

Cognitive and psychological biases are a process of our evolution: Our minds take these paths as a result of wanting to process information quickly. There are over 180 types of biases and heuristics that have been named so far by psychologists, social scientists, statisticians, economists and mathematicians. Many of these biases have been the result of work by the Nobel prize–winning

Daniel Kahneman, an economist, psychologist, professor and author, along with other great thinkers like Amos Tversky.

There are arguments that the judgements our unconscious minds make, made in milliseconds and based on limited data, can actually be more accurate than more deeply thought-out analysis – a claim that has become popular amongst some followers of the writer Malcolm Gladwell (2006). The following examples are five of my favourites that I have seen show up in my work with companies on data integration projects. These are all pertinent to leaders and managers implementing AI but are by no means an exhaustive list. Some of these you may already be aware of – in each instance, asking how this bias may show up in your team's or your own approach to implementing AI will help you understand how the bias may limit performance if not addressed.

Availability Bias

This is the reliance on just the information at hand. When looking at AI tools, you may have a preference for one due to hearing about a friend raving about it. For example, hearing about an AI disaster in another company may lead to you overestimating the same thing occurring in your organisation. If part of your AI implementation is to use a GPT within your company, for all employees to rely on it solely as a source of truth would likely be as much of a mistake as relying solely on 'the internet' or the top results from a search engine for your sources.

Anchoring

The 'anchor effect', first discussed by Kahneman and Tversky in 1974, describes a reliance on the first piece of information we have as a reference point for all subsequent information and decision-making. We see this at play, for instance, in how a price discounted from what was anticipated can persuade us to buy. Within AI implementation, if your team members have 'anchored' on a particular approach to AI implementation, tools, or policies, are these still the best ones to use in light of the developments in AI over the past year? AI development is moving fast, and there may be reliance on initial anchors in how it should be adopted, or in the risks involved in further integration. It is healthy to review and challenge these frequently, avoiding the reliance on your first 'anchoring' interaction with AI.

Dunning–Kruger Effect

This is the tendency of a person with limited ability in a particular field to overestimate their own ability (see Kruger & Dunning 1999). It also makes an appearance where some researchers show that the opposite occurs and some experts underestimate their abilities or falsely believe that others have the same level of knowledge. In both cases, it is linked to an estimation of other people's knowledge in a given area. This phenomenon can be observed particularly when there is a new technology that a limited number of people have understood or adopted. This leads to a specific threat for AI implementation:

Leaders need to be sure that those who are advising are capable of informing the process correctly, rather than overstating their ability.

Loss Aversion Bias and Status Quo Bias

Whilst these two forms of bias are separate, you rarely see one without the other in the workplace. Loss aversion states that we place a disproportionate weighting on avoiding losing something, far more than the value we place on gaining something. The popular book *Nudge* (2008), by the University of Chicago economist and Nobel Laureate Richard H. Thaler and Harvard Law School Professor Cass R. Sunstein, has some excellent examples of how they have demonstrated loss aversion in experiments repeated many times over with the same outcome.

Status quo bias is connected to this. Embracing change inherently involves embracing risk, which can make individuals hesitant to venture into uncertain outcomes. Not wanting to move from existing systems and processes due to wanting to maintain the status quo, or due to fear of loss, is a real warning sign as it can certainly harm change management and adoption.

Groupthink Bias

This form of bias relates to the social norms of the groups we belong to (see Janis 1972). The desire for harmony leads to conformity within the group, and we see people set aside their own personal beliefs and go along with the opinion of the rest of the group. This is an inherent danger in most workplaces that harms growth, and will certainly harm AI integration if not addressed.

How Human Bias Spreads

We've looked at how leaders should think of bias on two levels. Firstly, there are the cognitive biases that may be at play within your organisation that could impact integration and adoption of AI. Secondly, bias can creep into the AI systems themselves, as we've seen with examples earlier in the chapter. In cases where bias shows up, the algorithm is either executing data that it had been fed, or the algorithm itself is being informed by a human with inbuilt bias.

The following list of steps are broad suggestions on dealing with bias in AI, and each of these steps requires human/machine interplay.

Ensure Training Data are Diverse: AI models should be trained on diverse and representative datasets to minimise biases. This helps the model learn from a wide range of data points, reducing the risk of over-reliance on a specific subset of data. There could be a necessary education or awareness-raising exercise in advance of aggregating data for model training to ensure all parties understand what representative data looks like. Feature selection (choosing a subset of relevant features to enhance model selection) is key in reducing bias.

Blind Review: In situations where AI assists in diagnoses, having experts review some cases without AI input initially, and then with it to compare results. This helps identify any disparities between human and AI diagnoses.

Developing Feedback Loops: Implement systems where professionals or those that will use or be impacted by the algorithm or output can provide feedback on AI recommendations, and use this feedback to continuously refine the system. If the professionals on which you rely are not close to your end customer, can you be sure they really know what the customer base needs, wants or even looks like? Involving the people who will either use the system or be impacted by it could be a very useful part of development, and part of ensuring that the AI model remains accurate and unbiased over time.

Documentation and Guidelines: Developing guidelines and procedures that identify and mitigate potential bias in datasets. Documenting cases of bias as they occur, outlining how the bias was found, and communicating these issues can help ensure past instances of bias are not repeated.

Human-in-the-Loop: Another way to start resolving bias in AI is to have humans-in-the-loop to actively identify patterns of unintended bias. This reduces flaws in the system, creating a more neutral model.

Continuous Monitoring: Implement systems that continuously monitor the performance of AI models with fresh data and be ready to deploy newer versions if needed. This ongoing monitoring helps detect and correct biases that may arise during model operation.

Ultimately, the diversity of the team may be the ultimate tool in reducing bias. A diverse team can provide different perspectives and insights, which mitigates bias by generating more well-rounded and representative datasets.

Summary

In the shift to implementing AI in your company, the technical implementation of a system requires talent and expertise. The expertise, though, is beyond just implementing a product or process. Risk management is a major consideration. The risks associated with AI include how it will impact current jobs and the resulting effect on company culture, the impact on customer relationships, and the potential for 'rogue' behaviours when AI is not properly controlled. At the same time, there is the risk of moving too slowly due to being excessively risk-averse.

There are going to be differences of opinions between teams over the ethics and understanding of risk that the leader needs to deal with, and a need for alignment of teams who may have different stances around risk. For example, there may be differences of opinions over slowing down to achieve Quality Assurance (QA) of a product versus the desire to move fast. Having competitive activity and speed of movement as your only motivator is not wise. The leader needs to carefully navigate different points of view bought about by levels of comfort due to life experience. Within the team, there may be those who are digital natives and adept at using AI to increase efficiency, as well as people with grey hairs and battle scars from PR disasters which younger workers haven't yet had the misfortune of dealing with.

All of these experiences are valid, and need to be brought together to generate successful change and collaboration around AI.

GPT Prompts

- Act as a strategic consultant preparing a board-level report for an organisation initiating AI transformation. Outline the key considerations for data collection, storage, and governance to support scalable and trustworthy AI deployment. Address issues such as data quality, architecture, compliance, ownership, and cross-functional coordination. Structure the report to inform executive decision-making, risk assessment, and investment planning.

- Design a step-by-step operational plan for a company that supplies AI-enabled solutions to a governmental body within the European Union. List the specific technical, legal, and procedural measures required to ensure data privacy, security, and regulatory compliance under GDPR and related AI governance frameworks. The plan should be practical, auditable, and aligned with public sector procurement expectations.

What Questions Should a Leader Be Asking?

- What potential risks and challenges do we foresee in AI implementation, and how will we mitigate them?

- What processes are in place to control decision-making when using AI? Can we trace decisions back to their algorithmic processes?

- What measures are in place for monitoring and addressing biases or unintended consequences in AI outputs?

- What data security measures are in place to safeguard sensitive information?

- How do we ensure ongoing ethical review as AI technologies and applications evolve?

11

COLLABORATION AND CHANGE MANAGEMENT AROUND AI

Leading the team of the future

I watch the ripples change their size
But never leave the stream of warm impermanence
And so the days float through my eyes
But still the days seem the same
And these children that you spit on
As they try to change their worlds
They're immune to your consultations
They're quite aware of what they're going through

– DAVID BOWIE

In Chapter 5, we learned that the figure in 2016 for failed digital transformation was 70% according to research from the management consultancy McKinsey and Co. An update in 2021 revealed that 69% of digital transformations were failing (see De Smet 2023 and Engineering.com 2023). In this five-year period, why has not much moved forwards? Any leader who has led a company through a change knows all too well how hard it is.

In this chapter, we will discuss change management as it relates to integration of AI into your company's day-to-day practices and how to build collaboration through the change process. We will explore the practical steps for managing through change, and walk through an easy-to-follow framework that I personally use when coaching teams through change.

The Good, the Bad and the Ugly of Digital Transformation

One of the companies that is a poster child in the UK for successfully navigating digital change is RELX PLC, formerly Reed Elsevier. Reed Elsevier was predominately a publisher of business magazines and journals, founded in 1993 as a joint operation between two companies that had been operating since the late 1800s.

The shift to digital for large, old companies was predicted to be a challenge. Such a challenge, in fact, that in 1995, *Forbes* predicted that they would be the first corporate casualty of the internet. The strategy to change their business model to digital first and pivot away from print was already well underway by then. Their application of data totally changed their revenue picture. From print being the overwhelming majority of revenue, it fell to half of all revenue within a decade, and today it is less than 10%. Yet its market capitalisation during this period has quadrupled over the last decade to over $64bn, 8x higher than when the original merger went ahead.

I was struck by how this quote from Nick Luff, CFO for RELX, captures the longer-term view they took for managing change. 'I've done an awful lot of M&A [Mergers and Acquisitions], and what I've learned is the best thing you can do is drive organic development. It can't be done in a year. But don't underestimate the ability of an organisation to make a dramatic change over a three, four or five-year period, when you set it going in the right direction and keep driving at it, keep making small changes' (Management Today 2019).

During my time in this space, there have been many stories of digital transformation not working out quite so well. One story perhaps better known to people in the UK was of the government spending £10bn on a failed IT system for the National Health Service to manage patient records. Started in 2002, the National Programme for IT (NPFIT) was abandoned ten years later, leaving the UK taxpayer fuming at having to foot the bill for the failed transformation.

In the enquiries into the failure (see Henrico Dolfing 2019), the main points for the failure were highlighted as:

• A lack of clear leadership
• Not knowing, or continually changing, the aim of the project
• Not committing the necessary budget from the outset
• Not providing training
• A lack of concern for privacy issues
• No exit plans or alternatives
• A lack of project management skills
• Treasury emphasis on price over quality
• IT suppliers that depend on lowballing for contracts and charge heavily for variations to poorly written specifications.

Whilst an inability to manage contracts was a factor in this instance, inability to implement change was the biggest. In fact, in the many case studies I've seen that focus on failures of digital transformation, failures in the management of change always come up near the top of the list as one of the reasons. Yet I still see the same mistake being made time and again of not focusing enough on how to manage change.

The People At the Core

In my experience of researching the barriers that exist to successful change, whilst there are many factors that can affect different companies, the common denominator across all failed change projects is *people*. If you can't change your people, you can't change anything. Businesses like to underline the importance of 'strategy', but this word can have different connotations (even depending on whom you ask in the organisation). If we describe a strategy as the plan to achieve goals, a change initiative or change management process would be the vehicle that delivers that strategy. As change is hard, having a systematic approach to change allows you to bridge the gap between strategy and execution in the most seamless manner possible. The process involved in change and frameworks to follow are available, but process in itself is not enough.

People are notoriously difficult to change, but by understanding their motivations and following the steps that prepare people psychologically as well as the physical elements of the change, organisations can avoid becoming part of the future statistics of failed AI transformations. To avoid following that same failure rate requires adopting a different approach.

Understanding people in your company and what they will respond to is a vital step prior to implementing a change management process that is too often overlooked. As most modern leaders realise, the days of giving a command and having people obediently obey it are long gone (if they ever actually existed!). Even understanding the implications of the word 'change' is important to recognise – it's a default negative term, because it's associated with externally imposed change. Internally imposed change is generally referred to as growth or development. Resistance to change is a function of psychological safety and empowerment. If a person feels that the change is internally led or that it's in their best interests, it's growth. If they feel it's externally led or against their best interests, it's just change – which can be perceived as a threat.

As we take into account our learnings so far on the motivators for different generations, we know that a different consideration is needed for people of different generations in order to inspire the best output from them. As we apply a process for change management, we need to closely consider how to best integrate the different groups into the process rather than applying a 'one-size-fits-all' approach.

Bringing Change to AI: An Intuitive Framework

There are many views on how change around AI can be best achieved, and great management thinkers have spent many years researching models and frameworks that are as applicable for AI as they have been for other change management needs. Of the models that I have used to help teams manage change, I have found that following an adaptation of Kotter's

8-Step Process for Leading Change is one that clearly gives the best results. John Kotter is one of the leading experts worldwide on change management. His research has given businesses an excellent framework for navigating change. As times have progressed, however, there is an argument for an amendment to the last stage of the 8-step model, which outlines the stages necessary to affect successful change.[39] Kotter's original stages are:

1. Create urgency
2. Form a guiding coalition
3. Create a vision for the change
4. Communicate the vision
5. Remove obstacles to empower people to take action
6. Create opportunity for quick, short-term wins
7. Build on the change incrementally
8. Make it stick for the long term.

My gentle adjustment to the work of Kotter is to amend the last stage – 'Make it stick'. We all want to have initiatives that can last forever, but with the speed the world is changing, we are better off creating urgency (again).

We'll take a deeper look at these stages, but there are things to bear in mind upfront. Firstly, the path to bring change in AI will differ from organisation to organisation as no two

---

39 A deeper guide can be downloaded from https://www.kotterinc.com/methodology/8-steps/

change projects or cultures are the same. All eight stages should be followed in each case, but there may be greater emphasis on one stage or another, depending on the culture of the company. For instance, in organisational cultures with a hierarchical structure influenced by a national culture where hierarchy is respected, the guiding coalition may only be formed by well-informed senior leaders. In other organisational cultures, it may need people from all departments and levels. Change will occur best when it involves the people that can provide the biggest impact; this may mean that leaders need to be prepared to flout convention to ensure they get the best people for the job.

Upfront Failure

I've observed that when change has failed or slowed in an organisation, there's always one step in the eight that has either been skipped or not been given the right focus. Before using a framework such as this, the primary thing for a leader is to have a good understanding of the culture of the organisation. This will help them gauge the potential for change and identify barriers that may exist. There can be a large difference between the culture that the leader *thinks* exists, and the reality presented to the newer hires. You may already have a survey-based cultural assessment tool that gives you a sense of how all employees see the existing culture; if not, this can be a great way to get this information. Consider also using these kinds of surveys to assess propensity for change as well as attitudes around AI.

Also important before launching the stages is to ensure that the strategy for AI implementation is well-defined, aligned with company objectives and current company culture. Having a realistic timeline of what can be achieved before the process goes company-wide is essential. An unrealistic timeline will lead to frustrations both for you and the wider team, and when deadlines are missed, the whole process can fall apart as people get disheartened.

After gaining a clear picture of the culture and its readiness to change, and a well-defined strategy that aligns with company goals, the first three steps involve a core team centred around your leadership team.

Step 1: Create Urgency for This Team First

The wider company will follow your lead, but your leadership team is fundamental to the success of the project.

Step 2: Form a Guiding Coalition

Do you pick your eight best, or your best eight? The English universities Oxford and Cambridge have a sporting rivalry that dates back to 1829, where they row an eight-person boat down a four-mile stretch of the River Thames in London.

Competition for places in the boat is intense, with athletes spending years preparing themselves for that 15-minute all-out exertion. Mark de Rond, an ethnographer with Cambridge University's Judge Business School, spent time analysing the Cambridge crew's data to work out the best-performing team. What is fascinating is that it isn't the fastest eight individual rowers that make the fastest eight in the boat (Cambridge Ideas 2009).

Experience is essential – but experience in change around AI needs those with experience of AI as well as those with experience of managing change. In companies that I have advised on transformation, I always advocate for involving all levels. That doesn't mean that you need to slow the process by involving everyone at all levels, but having expertise from each level represented in the guiding coalition helps to ensure that unseen barriers don't later emerge.

As this team forms, be aware of any biases that may exist. We all have biases – these are generally learned processes that are a result of our inbuilt human survival mechanisms. For instance, it is less likely in most companies that the people with the most practical knowledge of AI are in the leadership team. Getting their expertise into the guiding coalition is vital. This may mean forming an unconventional coalition, with people in their 20s and 30s advising people in their 40s and 50s. If a biased view exists that age, tenure or traditional hierarchy will bring success, you will likely miss out on people with the best practical knowledge of AI application. It is necessary to make sure that this team can operate outside of the bounds of standard hierarchy if they are to flourish. Recognising biases will help you get to the best 8.

Step 3: Create a Vision for the Change, and Step 4: Communicate the Vision

Firstly, the vision needs to be created and communicated to your leadership team. There then needs to be a plan for how it will be communicated throughout the organisation. It is difficult to overcommunicate through change, which often fails as a result of absent or inaccurate communication. As a rule of thumb, take your existing plan for communicating through the change process, double it – and it will still probably be less than the amount needed.

Step 5: Empower People to Act by Removing Obstacles

The flow of information is important here – ensuring that you get clear advice on where barriers exist so that people can operate. You may not be able to remove every barrier immediately, but you need to ensure you can get access to unadulterated information on barriers that exist. Think through how clear the communication process in the company currently is, and how it can be altered to get the information you need.

Step 6: Create Opportunity for Quick Wins

Momentum is infectious, from the core team focusing on the guiding coalition through to the general employee that is being asked to embrace new working practices. All of these need to see that progress is being made. Celebrate these wins to build on the momentum.

Step 7: Build on Change Incrementally

Complex change, such as the change around AI, is a long-term process. It may feel like there is a contradiction between the need to move fast and the fact that successful change will take time, which is why communicating the successes of the short-term wins and building on these will help. As processes change and success is seen by using AI in one team, build on this by having this replicated in other teams. In addition, consider how frequently you will refresh the guiding coalition. The development in the AI world is evolving fast and showing no signs of slowing – if anything it is accelerating. Consider revolving people through the guiding coalition group to keep the knowledge and hunger sharp.

Step 8: Make it Stick for the Long Term. Then Create Urgency (Again)…

Summary

The adoption and use of AI and all the opportunity it brings will impact every team within the organisation, so participation from all groups is necessary. For the group to perform optimally, leaders need to harness the skills from their different life experiences, establish a base level of what AI *can* be used for, reconcile different interpretations of what it *should* be used for and introduce a decision-making process for how to implement it (ensuring everyone is informed).

The need for change should not be confused with a democratic process. As a consequence, a 'top down' process seems to be the default setting of most change management. However, there are differences you will see within your teams between people who will respond far better to being involved in the change process from an early stage as opposed to being on the receiving end of a company announcement of a new direction and an HR invite to attend training for how a new system needs to be used.

Following a tested framework for change such as Kotter's 8-step model gives the best chance of successful AI implementation. Ultimately, understanding the psychology of people's motivations and resistance to change will really help, especially if you can train your managers across the organisation to help their teams.[40]

---

40   See the Kubler-Ross Stages of Grief or Maurer's Resistance to Change models.

Top things leaders should watch out for include:

- Being unrealistic about what can be achieved and how long it will take to do it
- Not communicating effectively (either not enough, or not in the right way)
- Not identifying and addressing resistance
- Not involving the right expertise and seeking advice through the company.

Despite the complexity this change brings, it isn't something that can be avoided. In the words of John Kotter, 'Windows of opportunity are appearing more quickly than ever. Identifying an opportunity quickly and mobilizing urgency around it is key to beating competitors. Disrupt or be disrupted' (2019).

GPT Prompts

- Summarise the most commonly cited reasons for employee resistance to organisational change, drawing from the work of leading management thinkers such as Kotter, Lewin and Schein. For each reason, explain its root cause and provide a practical mitigation strategy that senior leaders can apply to maintain momentum, build trust, and reduce disruption during change initiatives.

- Provide a curated list of real or well-documented case studies where Kotter's 8-step change model was successfully implemented. For each case, briefly describe the organisation, the challenge faced, how the model was applied, and what measurable outcomes were achieved. Structure the examples to help a leader inspire sceptical or change-fatigued teams by showing what success can look like.

What Questions Should a Leader Be Asking?

- How will the process for change be set out, managed and assessed? What steps are needed to set the vision and get buy-in? What capabilities/perspectives/personality types do we need for our 'best 8'?

- How will we communicate the benefits and changes brought about by AI to our employees and stakeholders?

- How do we hear and understand people's concerns, and how do we work to overcome them?

- What collaboration is needed between different departments or teams to ensure successful AI implementation? How will this be instilled/assessed/maintained?

12

AI IMPLEMENTATION

Focusing Resources

*"Strategy without tactics is the slowest route
to victory. Tactics without strategy are the
noise before defeat."*

– Sun Tzu

D eveloping the strategy for how AI will be used within the business is essential before mobilising your organisation around it. In this chapter, we will look at AI investment and introduce a framework that outlines the stages of AI adoption within organisations. This can be used both to assess where your organisation currently is, and to help you identify future stages to develop your strategy towards. A balance must be struck to build strategy carefully, yet move quickly to seize the opportunity that AI offers and the benefit it can bring.

No Time For Standing Still[41]

Long before the words 'Artificial Intelligence' were being put into every sales pitch, I attended a session with Sir Martin Sorrell, who at the time led WPP PLC, the holding company for the advertising agency that employed me.

In the 15 years I had worked in advertising agencies, the media buying teams had separated themselves from being part of the 'full service' of advertising that was offered to clients into individual specialist companies. Sir Martin shared a reminder of the dangers of not innovating.

The day prior, he had assembled the heads of his creative and media buying agencies to watch a pitch for a start-up that was looking for investment. Their offering was a hugely reduced timescale in producing ads and placing them. The process of producing a piece of creative and having it appear in front of the right audience was months long. This pitch boasted that they would execute that process in 15 minutes. Sir Martin explained that when the company made this claim, there was a mixture of chuckles and incredulous scoffs of disbelief in the room.

The person leading the pitch went online and selected a piece of video for an entertainment company. With the use of their

---

41   In the words of legendary American Football coach Lou Holtz: 'Nothing on this earth is standing still. It's either growing or it's dying. No matter if it's a tree or a human being.'

editing tools, in the first 10 minutes they had cribbed together a 20-second video. At this stage, half of the faces in the room went white. He then proceeded to enter an online marketplace, where he found inventory that would reach the intended audience using a network of available websites, building a plan for how many sites the ads would appear on and how many times people would see them, taking a further five minutes. By this time, Sir Martin said, everyone's face in the room was white. What was taking the industry many weeks to do could clearly be accomplished in a much shorter time with the application of new technology. The message to us in the companies assembled was clear enough that we needed to adapt, but it was underlined by the statement: 'I'm investing in that company, so if that is the future, I'm going to be OK. I'm not so sure about you guys.'

Facing AI's Implications

Business history is littered with the remnants of companies that underestimated the change in front of them, and the need to not just innovate but to *keep* innovating. A successful investment in AI will involve far more than just seeing AI as a software 'product' that needs to be implemented. It requires company-wide coordination and alignment and may challenge some notions of the traditional decision-making hierarchy. For many companies, it will require a new way of thinking – moving away from the traditional, top-down 'command and control' approach, in which tenure in the company or seniority of position are the rationale for all decision-making. An area like AI may be so far out of the comfort zone of a senior leadership team that they opt for safety (and do nothing) for fear of the unknown. To do this may provide short-term comfort, but it is a big risk. The alternative is to plan for change, using the skills and knowledge in the company that are best suited to informing decision-making.

AI integration doesn't have to be 'all or nothing' – but the companies currently leading in this space are taking a specific view on both investment in AI and how learning in this area is working. One big development here is moving away from the traditional view where computers are taught how to perform by humans, and instead looking at how both parties can learn from each other.

AI Investment

Whilst some early adopter companies were ahead of the curve and have been building for years, it was the prominence of ChatGPT that brought it to life for the masses. AI suddenly started impacting many recruitment briefs and even events like Davos. Bosses are now demanding an AI strategy and in a rush to please, AI 'products' are being sold and the AI label is being stuck on some computerised systems that were not recognised as being Artificial Intelligence just a few months prior. Such is the need to avoid missing the bandwagon, the investment in AI is predicted to reach $632bn in 2028 (IDC 2024). 'AI Washing' is a very real thing to watch out for. This is the practice of claiming that a product is AI-based, even if it contains the smallest amount of generative technology – or to the unscrupulous, even less! Companies may end up paying more for a tool they believe has state-of-the-art AI capabilities which could yield inaccurate or incomplete data, leading to misinformed decisions.

One issue is the lack of consensus on a single definition for AI, and no downside to claiming that you have an AI product. However, with the amount of investment being raised in AI, this has become such an issue that regulators are stepping in (SEC 2024).

Efficient AI Implementation

The implementation of AI requires many leadership decisions. Perhaps the first question leaders ask themselves is the motivation for implementing AI at all. The top reasons given for implementing AI by business leaders include a mixture of defensive reasons such as a desire to obtain or sustain competitive advantage and a fear that competitors will use AI, or that new organisations will enter the market as a result of using AI. There are also opportunity-based rationales – including that AI will enable a move into new areas, and that customers will be better served with (or will expect) AI products (Ransbotham

2020). A reason often cited is the pressure to reduce costs requiring a shift to AI.

The next leadership question is often the costs involved so that the size of the investment can be understood. A chatbot can be built at relatively low cost. AI tools can cost from as little as $20 a month for a subscription to a GPT such as OpenAI's ChatGPT,[42] or for a generative AI tool for writing or images. The costs of implementing a system that you train yourself will be significantly more expensive: If you are looking to build a recommendation engine that rivals Netflix, the cost will likely run into the tens of millions of dollars. (As a reference point, OpenAI is reported to have spent $100m training GPT4.)

When deciding on costs, as well as the costs of tools and systems, there is the (more important) cost of investing in people in the organisation to develop themselves around AI. Some organisations look for benchmarks for how much others are investing. Many of the companies investing heavily in AI are technology companies. Whilst many of the businesses are in fields that may be different to yours, the following offers a look at the Nasdaq top companies' investment in R&D.

---

42   If you take the latest version of ChatGPT, prices are as of H1 2024 – older versions are available for free.

FIGURE 7

2024 R&D Spend

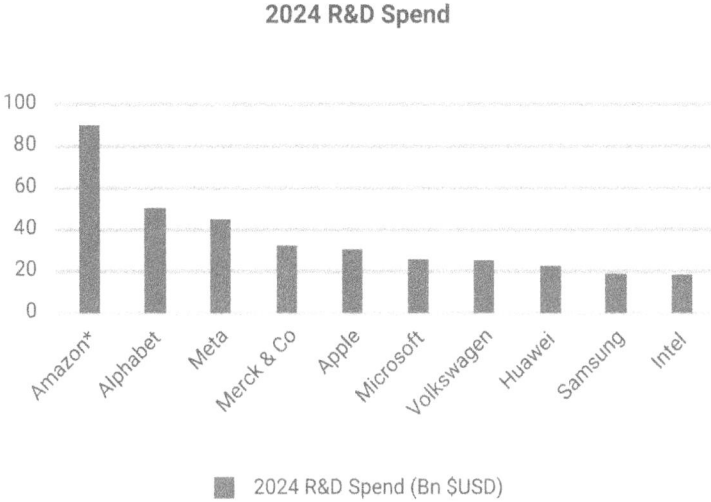

2024 R&D Spend (Bn $USD)

Sources: Annual company reports
*Amazon's listed figure is for all 'technology and infrastructure' spend due to accounting practices. All other companies depicted list R&D spend specifically.
**Apple's R&D spend reported as the fiscal year ending September '24
***Microsoft R&D spend reported as the fiscal year ending June'24

Whilst R&D will cover many things, it would be reasonable to assume that the companies best known for being leaders in AI are spending a good proportion of their R&D budget on it.

There are many factors that can influence the cost, which naturally is going to be connected to the complexity of the build and implementation. In addition to the upfront costs, there are associated ongoing needs that also carry a cost such as data format, data storage, data structure, data processing speed, the minimum accuracy rate for predictions, data visualisation needs and potential dashboard requirements.

Ensuring you are clear on the outcome that the business needs and that the strategy is placed in service of this before tactics are talked about is vital. Too often, these two elements of strategy and tactics get conflated, but they must be treated as distinct. Without this clear outcome and aligned strategic direction, it will be extremely difficult to get your teams behind an idea and measure its effectiveness.

When the outcomes are aligned, the tactics that will require focus are products and processes. A way to think about some of the physical elements of AI that may be a result of the desired output is as follows. The areas that many companies initially outline for AI investment may include:

- Chatbots or virtual assistants (conversational AI)
- Predictive analytics
- Recommendation systems
- Image recognition
- Fraud detection
- Natural language processing
- Personalised marketing
- Speech recognition
- Internal management systems.*

*The change of internal systems are sometimes a flashpoint where resistance is seen. You may see conflation between AI requests and 'automation needs'. In scenarios where the objective is cost-based (i.e. providing greater automation in order to generate greater margin), this is potentially what working teams resist the most.

Machines and Human Learning

The choice some leaders make as they think about applying AI is to follow traditional organisational thinking patterns. In hierarchical organisations, this can often follow a 'command and control' approach – applying AI as a product along very fixed lines, with a one-way application intended of using AI like any other software. A more sophisticated way of looking at AI is as an integration of what both parties can learn from each other. AI has no sense of hierarchy, departments or knowledge of what exists simply because a structure has been built around it. There is a gradient of how deeply AI can be integrated in decision-making and acting. For instance:

- Humans generate data/decisions, AI evaluates.
- AI is used for insight generation, humans use this info in a decision-making process.
- AI makes recommendations, humans decide what to do with those recommendations.
- AI decides what to do, humans implement it.
- AI decides what to do and implements it.

Those fixed in the traditional way of command and control thinking may go just as far with using AI to evaluate a human decision. There are certainly benefits to be had from this. However, as the patterns of AI 'thinking' or reasoning may be very different from human patterns, cultivating a relationship between teams and AI where each learns from the output of the other would be an optimal place for organisations to land.

The 4 I's of AI Integration

Earlier in this book, we met the 4 S's of data strategy sophistication. Applying a similar structure to stages that are seen with AI integration, these are four stages of AI adoption. The following stages can be used as a lens to help benchmark where your organisation is on the scale, and to help you set out what is needed to progress beyond your current stage.

Within each stage, the characteristics are (i) how AI is being used; (ii) how AI is brought into the organisation; (iii) the current levels of machine interaction with people; (iv) how people's AI skills are developed; and (v) leadership and ownership of AI.

One caveat that leaders should be aware of upfront: As we look at the stages of AI integration, you should not leap into a blanket goal of AI adoption or progression. AI may not be relevant, or indeed needed, for all situations. Being able to determine when to adopt AI and when not to adopt it will be an important distinction that your organisation needs to make. When this is determined, AI has a role to perform and the stages to follow are relevant.

The 'Intrigued by AI' Stage

At this stage, AI is seen much like other pieces of software. No company-wide definitions of what AI should be used for exist and there is a misunderstanding of what it may be capable of, with the belief that it could be a magical solution, if only it could be captured.

How AI is currently being used: In any process where AI could automate a current people-based task and bring greater efficiency – for instance, by using GPTs as a way of enhancing search capabilities or operating more quickly.

How AI is brought into the organisation: Free tools that vary across different teams.

Machine/human interaction: Sporadic use of AI products within teams for efficiency purposes within existing workflow structures.

How People are developed: Internal skills-building for all teams is not yet present. AI skills applicable to data and engineering may exist, but development is not invested in. AI experience may be looked for in recruiting, but is not well-defined.

Leadership/Ownership: No clear ownership. AI exists as a term on strategy docs without a clear path or owners identified.

'Igniting'

Organisation-wide understanding and adoption: AI is used by some teams, and the company's view of managing elements like risk is recognised only reactively in relation to tools that are purchased. Company-wide awareness of free AI tools that different teams may be using is low.

How AI is currently being used: Generative AI is being used for creativity; AI is being used in some teams for reasoning or self-correction.

How AI is brought into the organisation: Products with AI capability are purchased and made available to teams. Investment in data structure, data access and data management has been made, providing foundations for consistency.

Machine/human interaction: Use of tools or AI products for generative purposes is closely controlled. Humans are teaching machines and checking the output of AI.

How people are Developed: Experience or education in AI-relevant fields is sought in recruitment rather than developed internally. The organisation relies on self-teaching by team members.

Leadership/Ownership: Most activity is reactive. Policies are issued in firefighting mode when risks are identified. Ownership is placed with a single team (often IT) regardless of the area of expertise required.

Implementing

Companies at this stage have recognised the benefits that AI brings, and have teams who are using it, but coordination of the products used is still low, with different tools being used by different teams. The uses of AI are progressing beyond using it as simple software to using it for learning, and human/machine interaction is starting to be a two-way process.

Organisation-wide understanding and adoption: Company-wide aligned understanding of the different forms of AI, what can be used and by whom. Policies for usage exist, reactively put in place as products are brought in. Steps may have been taken to audit and control use of company-funded tools, plus other tools that employees may have access to.

How AI is currently being used: AI is used beyond reasoning/creativity/self-correction for learning. Teams are using AI processes to learn patterns from data and learnings made are introduced into the cycle of improvement.

How AI is brought into the organisation: The need for AI is scoped by business unit, recommendations being made for appropriate tools with use cases for predicted impact and return on financial investment. Some coordination between new tools and existing data systems and capabilities.

Machine/human interaction: AI may make recommendations for humans to decide whether to implement. Humans may devise the solutions and ask AI to evaluate. Machines may learn autonomously, but control is still tight and any adoption of information requires human reasoning with little ability to learn from AI.

How people are developed: Space is given for teams deemed relevant to develop new AI skills. New skills generally bought in via recruitment rather than grown organically. Adoption of AI is assessed within staff evaluations.

Leadership/Ownership: Departmental owners identified, some coordination between practitioners. Considerations for ethical issues and future development is reactive. HR and legal teams sought for advice on areas of responsibility, change management and risk.

Intelligent Application

At this stage, which may be several years into the company's AI journey, the understanding of the potential for AI is well understood, and the business is structured to not just adopt it but to make decisions around it. Leaders have a good understanding of the change management required for all staff to thrive, and for the company to continue to grow with AI. There is also a good understanding of the implementational process: Rather than rushing to bring AI adoption to life, time is allowed to ensure that a quality assurance process has been followed, and bias and ethical issues have been examined by the right teams or experts.

Organisation-wide understanding and adoption: Company-wide adoption with a structured approach to development, risk management and ethics. Advanced techniques that involve human and software integration are applied to areas such as machine learning, predictive analytics, NLP, computer vision and automation.

How AI is currently being used: AI is seen as an essential for market advantage, enhancing capabilities to enable the organisation to move into new businesses. Leadership systematically embeds AI for relevant applications throughout the entire organisation. Extensive changes have been made to many processes, not just to use AI but in response to what is learned with AI.

How AI is brought into the organisation: Coordinated approach to AI capability and operational development across the company. Integration between existing data systems and new processes is planned in advance of implementation with use cases identified and aligned across all teams. Alignment of different parties in the process of building with AI to allow time for quality assurance. Oversight group (possibly external) used to advise on risk and ethical issues.

Machine/human interaction: Humans and machines learn from each other and the relationship is beyond just teaching computers what humans know. Different adoption of decision-making based on situation: AI may make recommendations for humans to evaluate and implement or vice versa; alternatively, AI may decide what to do and implement.

How people are developed: Company-wide approach to development and change management of existing workforce to embrace AI usage with continuous plan for new development. Opportunities for people and AI to learn from each other is identified and built into learning loops. Managers developed as core identifiers of talent development and as change agents. Assessment of AI adoption is objective within performance evaluation.

Leadership/Ownership: Clear ownership, reporting progress at board level. All elements of integration to current systems, ethical and risk topics, change management and team development are guided in a coordinated manner by experienced experts in each area. An internal and potentially external oversight board developed.

A single-page overview of these stages is available via my website (www.onv.ai).

Key differences seen between stages include how companies think about controlling AI as opposed to working with it – moving from 'command and control' to enabling AI to be useful across the entire company. There is also greater coordination and alignment between 'back' and 'front' office (those responsible for building AI tools or integrating AI into processes and the end users). Companies succeeding with integration see that AI is not an all or nothing game – it is something that can be used to great effect where benefits can include time savings and cost efficiencies, but these are not the goals in isolation.

Summary

Investment in AI is essential. In the rush to say that something is happening with AI, some companies are investing in AI products without connecting it to the needs of the business, coordinating across the entire organisation, having a plan in place to help people develop the new skills they will need, or having managers trained in how to help their teams change.

As with any execution across your business, there needs to be a clear definition of the outcome desired, then defining the strategy – separating this from the tactics that operate in service of that strategy. The best plans have an evaluation phase, which helps to refine tactics to ensure continual learning. The 4 I's of AI Implementation should give you a guide to how developed your organisation is, especially in the thinking of how AI should be used.

Continuous improvement in AI comes from understanding the implementation of new technology and its capabilities and keeping up to date with developments that are coming at a breakneck speed. Most important to your business are the people who are going to use these tools – having the right people, building their skills, ensuring they work effectively together and then continually improve themselves. Having a culture in which people can be open about what they are doing, and can speak up about potential failure spots of AI that they can see without fear of recrimination, could very well prevent something going very wrong. This can only happen if there is a psychologically safe space in which people operate. All of this requires investment in your people. Next, we'll look at how to invest in those people.

GPT Prompts

• What mechanisms should a medium-sized business without a Chief Data Officer put in place for ongoing evaluation and improvement of AI systems?

• Explain how humans and AI systems learn to collaborate effectively in decision-making or operational environments. Include a progression from AI as a tool to AI as a teammate, and describe how trust, oversight, and human judgment evolve at each stage. Provide examples of how this progression is unfolding in real-world business contexts.

What Questions Should a Leader Be Asking?

- How are investment decisions for AI currently being made?

- How much of my budget is dedicated to R&D for AI, and how is that changing over time?

- How do we plan to stay updated on the latest developments and best practices in AI technology and applications?

- How do we propose to measure the return on investment (ROI) for our AI investments?

- What proportion of my development budget for AI is going on people?

13

INVESTING IN PEOPLE

The Most Important of the 3 P's

"A faster, smarter, cheaper Titanic is still a Titanic."

– ADE MCCORMACK

In the last chapter, we looked at the ways companies approach investing in AI. There are the immediately apparent costs – the costs of either hiring a company to build AI tools or the time cost of your engineering team building and running them. Then there are the costs associated with building the skills of your team. The product and processes you employ as tactics in the shift to AI adoption are nothing without people – the focus of this chapter. Of the three areas, this requires the biggest focus, yet as we've seen through data and digital transformations, not enough priority was put on helping people transform.

Successfully leading your current people through the change will require addressing concerns surrounding job displacement and fostering a positive attitude toward AI technologies. Education initiatives need to be established to ensure that employees are equipped with the necessary skills and knowledge to navigate the changing landscape effectively. In order to help your teams change and thrive in this new era, they need to be motivated by the return the role gives them. As we saw in Chapter 6, there are higher- and lower-order needs that all of us are looking to have satisfied via our roles at work, and as we've seen in Chapters 7 and 8, there are different motivators for the different age groups in the workplace. This investment in people is not as simple as buying in talent or running a training course. Instead, it will be about providing an environment in which people can work most effectively as well as how they communicate and collaborate.

The skills your organisation needs in AI may be developed by existing people reskilling or upskilling, or you may need to bring new skills in. Hiring to bring in people with AI skills has financial and cultural implications that will need to be managed. In this chapter, we will look at how to fill skill gaps as well as the focus you will need on your managers to develop skills to navigate these gaps. If you buy an AI product or try to implement a process to incorporate AI in existing activities without focusing on people, you may still end up with a faster, smarter, cheaper Titanic.

People at the Heart of Your AI Transformation

Businesses are based on relationships, and relationships are based on people (see Lemonis n.d.). Whilst we are born to be very selfish creatures, evolution has taught us that we are better off in groups to deliver our needs. The dynamics of groups are complex, but the motivation for healthy participation in any group comes down to our individual needs. Successful AI adoption needs skills from many different areas. It is about people working with machines, building skills to harness their power. For the foreseeable future, this will be the model in the vast majority of businesses: AI isn't about to grow human-level intelligence that we will feel the benefit of in all businesses, especially not in the short term. Whilst there are some people who forecast that the rate of progression may mean that replicating the complexity of the human brain is eventually inevitable, the processing power and subsequent cost is likely out of the reach of the average business for some time. Even if greater processing power became available and AI capabilities matched the dreams of the AI enthusiasts, before we get there, all of the non-engineering elements – such as governance to prevent AI destroying your brand – need your teams. This requires leaders building an understanding and response to these people's differing needs if you want to motivate them to work together to successfully integrate this new way of working.

223

There are three key groups leaders need to consider with regard to developing people in the organisation in relation to AI: (i) investing in current people; (ii) investing in new people; (iii) investing specifically in training managers.

Upskilling, Reskilling or Recruiting New Skills?

It is the competence gap more than AI itself that is the disrupter faced by people in organisations. There is the now-iconic quote from Richard Baldwin, of the Geneva Graduate Institute in Switzerland, that many people have taken to heart: 'AI won't take your job, it is somebody using AI that will take your job.' Your existing teams require the first focus in terms of development – the initial assessment will be what transferable skills already exist in the organisation that can be translated into working with AI without the need for external hiring. Whilst upskilling/reskilling is a lower-cost option than hiring anew, it still requires invest-ment –both in the change management process and in training people in new areas. It will also be the opportunity cost of the time taken to train/upskill/retrain your teams.

It is a fact of life that some people will not want to change, and this may mean that they part company with your business. The good news is that many people see the opportunity of AI. The 2024 annual workplace survey of 27,000 workers across 34 countries by Randstad (2024), a recruitment specialist and tal-ent consultancy, saw AI skills at the top of the list of those that employees want to develop, along with IT and tech literacy.

GenZers stand out above other age groups with their desire to develop AI skills. This may be due to their relationship with technology or knowing that they have up to four decades of professional life ahead to look forward to. 37% of the GenZ Randstad survey participants were acquiring skills and knowledge on their own. Only 10% received support from their company in terms of an organisational learning program, or implementation or cooperation with AI solutions. Nearly a third (29%) stated that they would even go as far as quitting a job that didn't offer adequate learning and development opportunities.

What Workers Want

As I entered the workplace in the 1990s as a Gen Xer, the advice of my Baby Boomer parents was to work hard for a company that would reward me for being loyal for 20–30 years. This was the path that the Baby Boomer generation in the UK had followed – a profession and a job for life were very real prospects for many to aspire to, followed by a final salary pension scheme, and most likely a wristwatch on departure. As my young mentor Nick enters the workplace as a GenZer, he is looking for reward for committing to 20–30 *months.*

Sometimes derided by the press as overly demanding with their expectations from the workplace, it is worth investigating more closely the things that a GenZer hopes to find within a salary package. Among the top things that GenZ look for in an employer are their attitudes to social responsibility, and taking mental health and wellbeing seriously (Dunlop et al 2023, Deloitte 2023).

In terms of developing skills themselves, the Randstad Workmonitor study showed wellbeing and mindfulness as one of the top three learning opportunities GenZ were most interested in, directly following AI skills and IT/Tech literacy (and ahead of communication and presentation skills and management/leadership skills).

The older generations certainly have to be thankful for the focus Millennials and GenZers have placed on mental health. No longer a taboo topic, seen as a sign of weakness or locked away behind a stiff upper lip, the open conversation about mental health has paved the way for people to be more productive – better focused whilst at work, taking fewer sick days and having greater output. This is not only better for us all from a humane point of view, but financially better for organisations. Research from the consultancy company McKinsey finds that employee disengagement and attrition could cost a median-size S&P company between $228 million and $355 million a year in lost productivity (De Smet et al 2023). Research by MHI and Business in the Community showed that the UK economic value of improved employee wellbeing could be between £130 billion to £370 billion per year. That's the equivalent of £4,000 to £12,000 per UK employee (Business in the Community 2023).

Understanding Your Company's Skills Gap

The starting point for most organisations will be to conduct an assessment of the emerging skills gap driven by the opportunity AI presents. Included in this should be an analysis of how recently transformative practices in areas such as automation and data analytics have fared within your organisation. If these areas were successfully developed within the organisation, was it a result of reskilling existing teams or hiring people with new skills? This may help you assess how the people in the organisation will respond to reskilling and the change needed in the AI transformation.

Recruiting for New Skills

In a Deloitte survey with 2,835 business and technology leaders involved in piloting or implementing generative AI in their organisations, over 70% said that they expect to make changes to their hiring strategy in the next two years because of AI (Deloitte 2024).

There are a host of new roles that have emerged already as organisations look to bring in talent. Roles that were unknown to many as little as a year ago such as prompt engineer, AI auditor and AI ethics manager are emerging, requiring specialised knowledge and expertise. As employers look to bring in AI skills, there is a balance to be struck between the value of these skills versus other relevant experience. A survey of hiring decision-makers in January 2024 by Reusmetemplates.com, a provider of CV templates, saw 56% saying that they would prefer to hire a candidate with fewer years of relevant work experience but more experience in AI. 10% of hiring managers surveyed said they would choose a candidate with far less relevant work experience but expert AI skills (ResumeTemplates 2024).

This may be influenced by the current need to recruit AI skills quickly, but placing these skills above other experience may come to have an impact on the culture of the company.

AI requires people across diverse areas, from data collection and integration of AI through to development of generative AI as well as business-specific roles, The title of 'Generative AI specialist' or 'ChatGPT specialist' is too broad for many requirements, and recruitment ads can now be seen for roles such as AI ethicist, chatbot content writer, security and privacy architect, and AI personality designer (!), alongside slightly more conventional analyst, engineering and data science roles.

Whilst these new roles currently exist, the ability for any new joiner to develop over coming years is really important. What is happening now in AI will look very different in five years' time. Leaders would do well to study how candidates have acquired skills. Those expanding their AI skill base through self-development and a judiciously chosen portfolio of roles will be most valuable in the adaptation they will need to do moving forward. The irony is that the AI-based ATS systems[43] for resume review currently at play are not trained to look for this kind of skill when they auto-screen a CV.

What to Remember When Hiring New Talent

Leaders need to consider that recruiting is not just about attracting people with the right salary, but also about expectations new joiners may have of the working environment. Recent years have seen a significant shift in expectations of working locations and conditions, and this is especially the case for younger generations. When looking to attract the best AI talent, this is a challenge that leaders need to address: If you find the skills you need among people who demand flexible working locations or working patterns, can this work within the existing organisational structure, or is the importance of employees being located in an office during work hours too high to sacrifice?

Managers – The Overlooked Element

Managers have a significant influence on the experience of people in any organisation, and in many cases even the lives

---

43 An Applicant Tracking System (ATS) automates the process of hiring including in many cases reading profiles or resumes for key words to ascertain suitability for a role.

of people who work in their teams. Yet it is staggering how few of us are actually taught managerial skills. Subjectively, in the circle of managers that I've engaged with, very few have had any formal training. Most managerial skills are learned 'on the job'; in some cases (like mine), a company may have put you on a two-day course 20 years ago. For any manager to have a successful, high-performing team, they need to understand the skills of management as well as understanding people and how to best relate to them. In the case of many managers who have been managing people for many years, they also need to understand how people have changed. In my case, since that managerial course two decades ago, people and their expectations from their roles in the workplace have changed a lot. A 25-year-old in the UK in 2024 is very different to the 25-year-old I was in the 1990s – we have had totally different life experiences that inform our view of the world.

Over time, many organisations have evolved their views of management in response to better understanding what helps people perform best, moving from the transactional form of management (motivating individuals via systems of reward/punishment for performing/not performing set roles) to seeing the benefits of transformational management. During the time I've been managing, a greater emphasis has been placed on the value of emotional intelligence.

Whilst emotional intelligence was starting to be talked about as I first started managing in the 1990s,[44] this was a topic that most managers had to discover for themselves. Emotional intelligence skills are of immense value in the modern workplace. The importance of this area is still underestimated by some managers and leaders who may be stuck within the older views of what leadership should be, as we saw in Chapter 4. In the requirement for the business to transform around AI, managers need to understand the different forms of motivation, and adapt their leadership approach to help them get the maximum out of each team member.

Managers in the AI age will need to be able to make their teams more adaptable, finding ways to empower the team to keep learning whilst making their world smaller. Successful managers will be those that encourage collective learning and development. Managers who can create versions of constantly learning teams within an organisation that is constantly learning will do well. The organisational heavyweight thinker Peter Senge has a great description of the learning organisation, which is one 'that continually enhances its capacity to create desired results, fosters new patterns of thinking, and encourages collective learning and development' (1994). If managers can deliver an environment for their teams to constantly learn, the organisation will certainly benefit.

---

44   Daniel Goleman's bestseller *Emotional intelligence: Why it can matter more than IQ* was first published in 1995.

As a leader, there are many skills you need your managers to have in order to navigate the integration of AI into your business. A mistake that is made in the selection of managers in too many companies is basing the selection on the individual being good at a particular role. However, the skills of management and the day-to-day functions required within the team are totally different. An assessment leaders need to make of their managers is how ready they are on the 'soft' skills front. As each business is unique, there will be a set of things specific to your business that is a minimum requirement for a manager to be successful. There are the minimum day-to-day expectations from the role you need from your managers, items such as building and implementing a strategy for their area of responsibility, reporting on progress, effective budget management, ability to assess performance and course correct where necessary. Then there are the less obvious 'softer' skills that are essential for managing a team, as they lead their teams through the changes that AI integration into your business will require. It is worth assessing where your managers are on softer skills that are relevant in modern teams, such as:

- How to operate a team of different life stages and experiences
- How to operate a team of different national cultural backgrounds
- How to understand the strengths of the team (both the members' individual strengths and the collective strength of the team)
- How to operate a team with contrasting complementary strengths

- How to assess their team members' skills and help them develop new skills
- How to seek and give feedback appropriately
- How to manage conflict effectively
- How to ensure effective communication between members of the team, and between their team and other teams in the organisation.

As the impact your managers have on the individuals is so important, leaders need to invest in building skills for this group. Manager development can be done in-house if the structure of your company allows for it, taking outside help wherever possible. My experience has been that developing an educational program with companies that specialise in these specific areas has the best impact on skills-building.

Manager development should also be considered as an ongoing requirement – as the role of managers has so much influence on the success of your organisation, and on the organisational experience of the people who report to them, this group is too valuable to not constantly invest in improving.

Different Communication Preferences

Establishing the best methods of effective communication across all people in the organisation plus ensuring that these are aligned across all teams is vital, yet this is a common failure point for many organisations. Across the different organisations and people I've worked with, there have been different preferences for communication. I've worked with clients whose strong preference is long-form written communication in emails and memos. I've also worked with people where brevity is everything and short-form, message-based communication is highly preferred.

The use of social media–inspired workplace tools in which employees 'post' information has been incredibly successful, as these tools replicate a format that people have become fond of using to communicate with friends. There are also team collaboration workflow tools, internal websites, newsletters, company noticeboards and company-wide meetings or 'all hands'.

There are benefits and drawbacks to each. The important point is that whatever method is chosen must be (i) fit for the purpose required and (ii) used by everyone throughout the organisation. The method used should be efficient and easy for communication to be made and received by anyone who needs it. It needs to be searchable with minimal effort and, vitally, should be easy for all groups of employees to use.

These methods should be reviewed periodically – especially taking into account the views of the newest entrants to the workplace. The youngest members of teams have very different experiences of how information can be shared efficiently compared to someone whose workplace experience included using fax machines to convey information.

Having a policy around which communication channels should be used to avoid new channels of communication popping up is a smart move many organisations have made. There's nothing worse than having to check email, Messenger, Teams, Asana, Slack, the internal intranet, your company's social media page, newsletters and company blogs to try to find the information you need.

The importance of establishing clear communication for all is too often overlooked and its importance can't be overstated. In the words of George Bernard Shaw, the Nobel Prize–winning playwright, 'The single biggest problem in communication is the illusion that it has taken place.'

What Steps Must a Leader Take When Building a Great Team For the AI Generation?

Investment in people for AI is not just about the hiring process, but a systematic approach to improvement of people throughout the organisation. The starting point will be to conduct an assessment of the emerging skills gap driven by the opportunity AI presents for your organisation. Within this initial assessment should be a view of what transferable skills already exist in the organisation that can be translated into working with AI without the need for external hiring. In addition, leaders should look for analysis of how recently transformative practices in areas such as automation and data analytics have fared within their organisation. If these areas were successfully developed within the organisation, was it a result of reskilling existing teams or hiring in new skills? This may help you as you develop the strategy for acquiring people with the right AI skills.

As you establish the skill gaps, you need to bear in mind that the differing age groups in their organisations will have different motivating factors if you want to attract and/or retain them: Recruiting, motivating and developing someone who is from GenZ is different from doing this with someone from the Baby Boomer generation.

Your managers are a key group that you must invest in. With all groups, as AI continues to develop, you need to continue to invest to keep up with the changes that AI will continue to bring and to continually improve.

Bringing the different teams together into a harmonious unit requires people developing an understanding of each other. Teams who understand each other will operate way better, and systems of communication between people are at the heart of this. Consider reverse mentoring either one-on-one or in wider forums to help your employees understand how different age groups work. Shadow boards have become an approach used successfully by some companies (Jordan & Sorell 2019). Facilitated offsite exercises can also work wonders to bridge the gaps/preconceptions between teams.

Finally, as with all areas of the business, a process is needed for evaluating the investment in people that is going to be made, bearing in mind that a long-term, not short-term, view is necessary, and may require different models of evaluation than are used for other investments that the organisation makes.

GPT Prompts

- List in order of effectiveness, as reported by users, the different methods of communication in organisations.

- What are the preferred tools of organisational communication for different generational groups?

- Advise me as a company leader on the communication tools that my teams like most and which they like least.

What Questions Should a Leader Be Asking?

- What is the skill gap between the current and desired state with regard to fulfilling the company's AI strategy?

- What is the budget reserved for the people strategy, and how does this compare to other AI investment areas?

- What is my organisation's process for evaluating the impact of investing in people relative to products or processes?

- What is the company doing to upskill managers to understand the needs and motivations of people from different groups in order to develop teams that can perform highly with AI?

CONCLUSION

Leading in an AI World

*"The past is a foreign country. They do things
differently there."*

– L.P. HARTLEY

Getting the Most From AI

Throughout this book, we have looked at the combination of practical elements involved in implementing AI. Leaders and their leadership teams need to start first with the base understanding of what AI is, what possibilities and limitations it has, and how to define its application in the context of their business. The strategy for AI must lead with the value that it brings to the organisation – aligning with the objectives of the business.

AI does bring a new set of parameters for a leader to consider. There are many steps needed – the ethical considerations, potential for risk, ensuring compliance, developing collaboration, evaluating the efforts and of course leading through the change management needed.

The stages of maturity of AI integration that we met in Chapter 12 are a starting point for looking at your organisation's current status across these six dimensions:

- Your organisation's current understanding and adoption of AI
- Current applications of AI within your organisation
- How AI is brought into the organisation
- Levels of and processes for machine/human interaction
- How your people are being developed with AI skills
- Overall leadership/ownership of AI.

For many organisations, AI will surpass even the placing of a computer on every desk in terms of the advantages it can bring. There are jobs that will be impacted by AI such as manufacturing, or roles that involve routine admin tasks such as data entry or document processing. Some jobs previously performed by humans may be replaced, while new jobs will be created.

Even though we have all been the recipients of enhancements from AI in our experiences as consumers, for many companies it was not something they actively thought about integrating into their daily operations until the advent of generative AI. It is pretty easy for companies to build generative AI pilots – or conversely to raise a policy prohibiting the use of AI products in the workplace. Turning this work into at-scale capabilities that capture the benefit AI can bring to your organisation is a deeper challenge.

Generative AI in particular has the potential to generate a huge advantage for organisations, impacting working hours through either automation or augmentation. This would free up people in your organisation to focus on higher-value activities. The figures touted by think tanks and consultancies of the economic output boosts that GenAI will bring are significant.

Fools Rush In Where Angels Fear to Tread[45]

The assessment of where your business currently stands helps establish your organisation's existing AI status. But before being tempted to build a program to move through the 4 I's of AI integration, leaders must have at the forefront of their minds the fact that AI tools and applications can only be useful if they are in service of the objectives of the business. The importance of aligning your leadership team around the ground truth of applying AI in service of your business objectives cannot be overstated.

---

45   Pope, A. (1711).

Ensuring you are clear on the outcome that the business needs, and that the strategy is set in service of this before tactics for deployment are talked about, are therefore key steps. As all leaders know, the best strategies are well thought through, weighed up on the balance of where resource and expertise are available. In the rush to get an 'AI strategy' and not be left behind, beware of the many companies that will happily try to persuade you with some sharp marketing that the system they offer will solve all of your AI dreams. Defining what AI is and isn't, and what it needs to do *in the context of your business* before adopting any tactics to deploy AI, will help avoid wasted projects and failed change management programs.

Your leadership team will have witnessed or may even have been part of failed digital transformations: Apply the benefit of these learnings to prevent the same thing happening to your organisation, which would waste time and money in failed AI integrations.

Taking advantage of AI to progress beyond your current position will depend on the ability to successfully navigate the application of AI products, and to lead your organisation though the change. How much your organisation can benefit will depend on people – and your ability to bring the right people with the right skills and the right hunger to navigate these challenges and answer myriad tricky questions.

Evolving Our Understanding of People

As we saw in Chapter 13, people are the most important part of the equation of successful AI integration – more so than the products you can use or the processes you can put in place. AI tools are nothing without the human expertise to refine and implement these tools effectively. Successful alignment of people is perhaps therefore the most important role in the entire business that a leader can focus on.

As we have learned from unpacking some of the challenges of transformation that slowed digital transformation down (and where many companies still find themselves unable to progress), the challenges with change come down to people. To bring the power of AI to life within a company requires deep understanding from leaders of the potential advantages that this technology will bring, but also a deep understanding of the people in the organisation who are best placed to deliver it.

During the decades I have operated as a manager, watched other managers and helped managers develop, I've witnessed the approaches to managing people change over time. Fortunately, as we saw in Chapter 4, we have mostly moved on from the transactional approach that was popular in the 1960s to the 1980s. With the benefit of research and a fair amount of failure, and by adapting to the changing expectations that each generation has brought, we've come to understand that modern management and leadership requires an in-depth understanding of people.

It was this that led me to learn about the power of EQ in management and ultimately to return to university in my 40s to study organisational psychology. Advances in understanding EQ, for instance, have brought leaders higher levels of staff retention and greater workplace happiness. To deny the link between happy employees and productive employees would be reserved for either the most cold-hearted, or perhaps foolish, of leaders.

We are all individuals with hopes, dreams and desires alongside needing to have our basic needs met. The 'hierarchy of needs' that Abraham Maslow defined (1943) provides a framing that has stood the test of time well. Our lower-order needs are fulfilled through our participation in social groups. There are basic needs – from physiological ones through to things like safety, security and belonging. But our workplaces can provide the answer to our higher-order needs that we see towards the top of the pyramid that is popularly used to depict his model. Self-actualisation – the ability for us to realise our potential – and acceptance are examples of the higher levels of psychological development that can be delivered via the workplace. Successful leaders tap into the motivations associated with having our own individual needs met via our workplace experience.

No matter who we are, these needs, from basic to advanced, are universal. The TV chat show host, author and powerhouse Oprah Winfrey said that in 35,000 interviews with people from political and business leaders, she was asked one question almost every time at the end of the interview: 'Was that ok?'[46] We all need acceptance. The satisfaction brought by purpose and acceptance, and the opportunity to achieve the hopes, dreams and desires that are core to our human experience, can motivate the very highest levels of delivery from your team.

The Rich Tapestry of Our Teams

There are many social influences on our development – an important one being the national culture that we are shaped by in our formative years. As we have seen as we looked at how groups are formed in Chapter 6, the work of academics Geert Hofstede and Erin Meyer have helped us understand the impact of national culture on our outlook, and provided groundbreaking tools that can be employed in organisations to help better understand the people in our teams.

The globalised nature of the modern workplace, and work that has been started to correct inequality in the workplace, encourages a far more diverse set of people in our workplaces than many members of the Baby Boomer and Gen X may have seen in the early parts of their careers. This diversity of experience and values brings invaluable perspective to the workplace: a

---

46 Oprah Winfrey, Harvard 2013 Commencement Speech: 'I've done over 35,000 interviews in my career. And as soon as that camera shuts off, everyone always turns to me and inevitably, in their own way, asks this question: "Was that OK?"'

deeper and richer ability for organisations to solve problems and make better decisions.[47]

As we have seen throughout this book, the element of diversity leaders must focus on for successful AI implementation is the age-related diversity brought by involving different generations in the strategy, change process and deployment of AI. The two key groups of people in your organisation that require the most focus are the ones at opposite ends of the age/career spectrum: the longest-standing members of the workforce and the newest entrants. Of course, all the different generational groups across the workforce are going to be vital to have on board for successful adoption of AI, but the focus in this book on the newest entrants (who have the smoothest path to adaptability to AI) and the people who are in positions of influence and/or control is deliberate. These groups will have an oversized influence on the successful implementation (or otherwise) of your AI strategy. Both their unique characteristics and approach to work can provide the perfect complementary set of skills to ensure the organisation can adapt to AI in the best way.

---

47   This intersection between national culture and age is becoming increasingly important in our diverse workplaces for leaders to be able to navigate. My own research and resulting thesis on this, *Leadership and Generations in the UAE*, is available on my website (www.onv.ai).

Leading Change

The landscapes within our organisations are under constant change due to technology, as much as our lives outside them are. The advances in technology that those of us from the Baby Boomers, Generation X and GenY have felt during their careers may well cause us to feel that this change is accelerating.

Keeping up with change may feel exhausting, but history is littered with companies that failed to do so. More than half of the companies in the Fortune 500 list in 2003 were no longer in the list in 2023. In the FTSE 100, since 1984, 72% of the original companies have disappeared from the list. There are many reasons why companies may fail, and topping the list of reasons is certainly a failure to change and reinvent to stay ahead.

The move to embrace AI for organisations requires leaders to focus above all else on helping people through the change process. People can be resistant to change – this is a topic much studied by psychologists and sociologists. Fortunately, through these studies we are better equipped these days to understand the cause of resistance, and to have well-researched models of change management, such as Kotter's Stages of Change that we met in Chapter 11.

The change required is both of existing skills and practices and a change in assumptions over who is best placed to lead this change. One of the steps in the Kotter framework for change (see Kotter & Cohen 2012) is to build a guiding coalition. The skills needed in the implementation of AI are owned by very different groups. Just reflecting on a few of the chapter-ending questions in this book in the area of technical implementation and risk management can help see how different skill sets need to work in harmony to answer them. Develop involvement in this guiding coalition of two key groups: the newest (youngest) entrants to the workforce and the longest-serving in their career (oldest) employees.

The Only Constant

On the way through this book, we have journeyed through how AI is currently defined and looked at applications of AI and the pitfalls that exist if AI is not properly controlled. We've looked at ethical challenges and the dangers of bias being inadvertently brought into an AI system. We've looked at the opportunity of learning from the last major landmark change in organisations – the digital transformation – to avoid repeating the same mistakes.

The successful integration of AI needs change. The philosophy of Heraclitus, that the only constant in life is change, is as relevant today as it was when he lived 2,500 years ago. In our modern workplaces, the greater application of the basics of psychology in the workplace has helped us to understand our organisations' greatest assets – our people – far better. To focus on people in organisations as merely replaceable 'resources' misses huge opportunities and leads to wasted efforts, costing organisations dearly in both time and money. Greater understanding of the needs and consequent motivations of individuals, as well as better acceptance of how we form and operate in groups, are the vital frontiers for organisations to master.

As change is constant, there is also a need to build capabilities within the company to keep up. The speed of development within this field is breathtaking. The output of AI is getting more and more impressive every day, technological tools for enhancing efficiency abound, and with the current investment in generative AI, these will undoubtedly improve. Crafting the vision to not just get up to speed with current forms of AI but also keep up as it evolves is essential.

Keeping AI in Perspective

The opportunity that AI brings may well be the most significant opportunity that leaders will see in their careers. It requires new thinking in order to take advantage of it. The job of leading in the AI age requires precisely the same approach that has served leaders well as they have built their organisations thus far – developing strategies to get to the required outcome by deploying resources.

Leadership of the organisation remains the same: bringing teams of people together behind a vision, motivating them, measuring the impact of resources deployed, course correcting, growing, and embracing new opportunities. All of this remains as necessary in the world of AI as ever before.

It must be remembered that whilst AI has the potential to be an incredibly powerful tool, it is not magical: Its capabilities and outputs are very much defined by the people who implement it and use it. The key to all of this is the people you put at the heart of this revolution within your organisation, and particularly the way you bring together the two vital groups that we have focused on in this book: the older generation who most likely occupy positions of influence, and the younger generation who have the best relationship with technology.

There are many challenges to be overcome and problems to be solved. Some leaders may see these issues as something they may be able to avoid by banning AI from being operated within their companies. You cannot stop the advancement of this new technology; what you can do instead is to ensure that the huge potential it carries can be applied to best effect. Implementing AI is not without risk. As we have seen, there are dangers that exist if AI is not carefully deployed – but these dangers need to be weighed up against the danger of not adapting to the world of AI.

In the words of the man who led the company where I proudly spent a quarter of my entire life, the biggest risk is not taking any risk.[48]

---

48   'The biggest risk is not taking any risk ... In a world that is changing really quickly, the only strategy that is guaranteed to fail is not taking risks.' Mark Zuckerberg (Tobak 2011).

Bio

Tony Evans

Tony is a tech veteran, team coach and organisational psychologist.

His career has taken him from working at leading ad agencies through to spending 13 years at Meta as it grew from 1,000 to over 80,000 employees.

Tony holds qualifications in marketing and advertising and gained an MSc in Business Psychology in 2022.

His research into national culture, generations and leadership is used to help companies structure themselves effectively to take advantage of the opportunities that AI brings.

Tony's approach is 'always people at the heart', placing the importance of focusing on people above products or processes in any organisational change.

Originally from Kent in the UK, Tony now resides in the UAE, proudly calling Dubai his home.

Works Cited

Introduction

Shane, J. (2024, February 10). DALL-E3 generates candy hearts. *AI Weirdness.* https://www.aiweirdness.com/dall-e3-generates-candy-hearts/

PART 1 – ADAPTING TO THE TIMES

Chapter 1: ADAPTING TO THE TIMES

Asch, S. E. (1951). Effects of group pressure upon the modification and distortion of judgments. In H. Guetzkow (Ed.), *Groups, leadership and men: Research in human relations* (pp. 177–190). Carnegie Press.

Hofstede, G. (2001). *Culture's consequences: Comparing values, behaviors, institutions, and organizations across nations.* Sage Publications.

Milgram, S. (1963). Behavioral study of obedience. *The Journal of Abnormal and Social Psychology, 67*(4), 371.

Zimbardo, P. G. (1971). *The power and pathology of imprisonment.* Congressional Record. (Serial No. 15, October 25, 1971).

Chapter 2: GENERATIONAL DIFFERENCES

Occupy movement. (2024, November 7). In *Wikipedia*. https://en.wikipedia.org/wiki/Occupy_movement

Taylor, A. (2011, October 17). Occupy Wall Street spreads worldwide. *The Atlantic*. https://www.theatlantic.com/photo/2011/10/occupy-wall-street-spreads-worldwide/100171/

Woods, A. (n.d.). The death of Moore's Law: What it means and what might fill the gap going forward. *CSAIL Alliances*. https://cap.csail.mit.edu/death-moores-law-what-it-means-and-what-might-fill-gap-going-forward#:~:text=There%20are%20other%20%E2%80%9Cexotic%20technologies,contender%20to%20replace%20Moore's%20Law

Chapter 3: AI IN THE WORKPLACE

Dyson, G. The Third Law. In J. Brockman (Ed.) (2019). *Possible minds: Twenty-five ways of looking at AI* (pp. 35-40). Penguin Press.

Hu, K. (2023, February 3). ChatGPT sets record for fastest-growing user base – Analyst note. *Reuters*. https://www.reuters.com/technology/chatgpt-sets-record-fastest-growing-user-base-analyst-note-2023-02-01/

Kissinger, H., Schmidt, E., and Huttenlocher, D. (2022). *The age of AI: And our human future*. Back Bay Books.

McCarthy, J. (2006, June 26). The philosophy of AI and the AI of philosophy. Stanford University Computer Science Department.

Shane, J. (2019). *You look like a thing and I love you: How artificial intelligence works and why it's making the world a weirder place*. Wildfire.

The CEO's guide to generative AI. (2023). IBM. https://www.ibm.com/thought-leadership/institute-business-value/en-us/report/ceo-generative-ai

Chapter 4: MODERN LEADERSHIP

Dastin, J. (2018, October 11). Insight – Amazon scraps secret AI recruiting tool that showed bias against women. *Reuters.* https://www.reuters.com/article/us-amazon-com-jobs-automation-insight/amazon-scraps-secret-ai-recruiting-tool-that-showed-bias-against-women-idUSKCN1MK08G/

Larson, J. et al (2016, May 23). How we analyzed the COMPAS recidivism algorithm. *Pro Publica.* https://www.propublica.org/article/how-we-analyzed-the-compas-recidivism-algorithm

Lightfoot, L. (2016, June 24). The student experience – then and now. *The Guardian.* https://www.theguardian.com/education/2016/jun/24/has-university-life-changed-student-experience-past-present-parents-vox-pops

Vartan, S. (2019, October 24). Racial bias found in a major health care risk algorithm. *Scientific American.* https://www.scientificamerican.com/article/racial-bias-found-in-a-major-health-care-risk-algorithm/

Chapter 5: THE GREAT EFFICIENCY ILLUSION

Abercrombie, PL. & Geddes, P. (1915). Prolegomena: On Professor Geddes' book – Cities in evolution. *The Town Planning Review*, *6*(2), 137–142.

Block, C. (2022, March 16). 12 reasons why your digital transformation will fail. *Forbes.* https://www.forbes.com/councils/

forbescoachescouncil/2022/03/16/12-reasons-your-digital-transformation-will-fail/

Bucy, M. et al. (2016, May 9). The 'how' of transformation. *McKinsey.* https://www.mckinsey.com/industries/retail/our-insights/the-how-of-transformation#/

Calaprice, A. (ed). *The Ultimate Quotable Einstein.* Princeton University Press.

Harvard Business Review. (2022, January). *Analytic services survey.*

Lamarre, E. et al. (2023). *Rewired: The McKinsey guide to outcompeting in the age of Digital and AI.* Wiley.

Rogers, E. (1995). *Diffusion of innovations.* Free Press.

Sherif, A. (2024, March 13). Digital transformation spending worldwide 2017–2027. Statista. https://www.statista.com/statistics/870924/worldwide-digital-transformation-market-size/

PART 2 – THE PEOPLE WE NEED

Chapter 6: GROUP FORMATION

Dembosky, A. (2012, February 3). Facebook millionaires eye new exploits. *Financial Times.* https://www.ft.com/content/dda53980-4d47-11e1-bdd1-00144feabdc0

Janik, V. M. et al (2006, May 23). Signature whistle shape conveys identity information to bottlenose dolphins. *Proceedings of the National Academy of Sciences,* *3*(21), pp. *8293–7*, https://doi.org/10.1073/pnas.0509918103

Meyer, E. (2016). *The culture map.* PublicAffairs.

Proceedings of the National Academy of Sciences. (2006, May 23), *103*(21).

Schein, E. H. (1985). Organizational culture and leadership: A dynamic view. Jossey-Bass.

Tajfel, H. et al. (1979). An integrative theory of intergroup conflict. In Mary Jo Hatch & Majken Schultz (Eds.), *Organizational identity: A reader.* Oxford, pp. 56–65.

Chapter 7: ARE THE BOOMERS OK?

Block, C. J. (2021). *Business is personal.* Passionpreneur Publishing.

Britannica, T. Editors of Encyclopaedia (2024, September 23). Y2K bug. *Encyclopedia Britannica.* https://www.britannica.com/technology/Y2K-bug

Know Your Meme (n.d.). OK Boomer. https://knowyourmeme.com/memes/ok-boomer

Removal of Sam Altman from OpenAI (2024, November 4). In *Wikipedia.* https://en.wikipedia.org/wiki/Removal_of_Sam_Altman_from_OpenAI#:~:text=On%20November%2017%2C%202023%2C%20at,consistently%20candid%20in%20his%20communications%22

Seligman, M. E. (1989). Research in clinical psychology: Why is there so much depression today? In I. S. Cohen (Ed.), The G. Stanley Hall lecture series (pp. 75–96). American Psychological Association. https://doi.org/10.1037/10090-006

SpencerStuart (2023). UK 2023 Spencer Stuart Board Index. https://www.spencerstuart.com/research-and-insight/uk-board-index

Chapter 8: THE KIDS ARE ALRIGHT

Costa, K. (2024). *The 100 trillion dollar wealth transfer: How the handover from Boomers to Gen Z will revolutionize capitalism.* Bloomsbury Continuum.

Deloitte (2023). The Deloitte global 2023 GenZ and Millennial survey. https://www2.deloitte.com/cn/en/pages/about-deloitte/articles/genzmillennialsurvey-2023.html

Hearn, E. (2019, November 7). *Boxing Roundtable: KSI Says 'Bring On Justin Bieber!'* [Video]. YouTube. https://www.youtube.com/watch?v=do9SSy_Eh4o

Lunn, T. (2021, September 24). CENTRE STAGE Anthony Joshua top five PPV buys: Andy Ruiz Jr may have been his biggest payday but viewers poured in to watch him fight Wladimir Klitschko and Joseph Parker. *Talksport.* https://talksport.com/sport/948078/anthony-joshuas-top-five-ppv-buys-all-time-andy-ruiz-jr-rematch-payday-usyk-fight/

OECD (2021). 'All the lonely people: Education and loneliness', Trends Shaping Education Spotlights, No. 23, OECD Publishing, Paris, https://doi.org/10.1787/23ac0e25-en

Peterson, B. (2023, June 16). What Gen Z wants in the workplace. *Washington Post.* https://www.washingtonpost.com/business/2023/06/16/gen-z-employment/

Randstad (2023). Randstad Workmonitor 2023. https://www.randstadusa.com/landing/workmonitor-2023/

Twenge, J. (2023). *Generations: The real differences between Gen Z, Millennials, Gen X, Boomers, and Silents–and what they mean for America's future.* Atria Books.

US Census Bureau (2023). Household Pulse Survey Public Use Files, Phases 1 and 3.10.

Chapter 9: BRIDGING THE GAP

Cattell, R. B. (1963). Theory of fluid and crystallized intelligence: A critical experiment. *Journal of Educational Psychology, 54*(1), 1–22. https://doi.org/10.1037/h0046743

Fokina, M. The future of chatbots: 80+ chatbot statistics for 2025. *Tidio.* https://www.tidio.com/blog/chatbot-statistics/

Gelles-Watnick. R. (2024, January 31). Americans' use of mobile technology and home broadband. *Pew Research Centre.* https://www.pewresearch.org/internet/2024/01/31/americans-use-of-mobile-technology-and-home-broadband/

Hendrix, J. (2024, January 15). Questioning OpenAI's non-profit status. *Tech Policy Press.* https://www.techpolicy.press/questioning-openais-nonprofit-status/

Mannheim, K. (1952). Das Problem der Generatione. In Paul Kecskemeti (Ed.), *Karl Mannheim: Essays.* Routledge, pp. 276–322.

Tversky, A. & Kahneman, D. (1974, September 27). Judgment under Uncertainty: Heuristics and Biases. *Science, New Series, 185*(4157), pp. 1124–1131.

Unlocking the secrets of search, by generation. (n.d.). *Fractl.* https://www.frac.tl/unlocking-secrets-search-generation/

Yang, M. (2023, September 22). The vast majority of NFTs are now worthless, new report shows. *The Guardian.* https://www.theguardian.com/technology/2023/sep/22/nfts-worthless-price

PART 3 – IMPLEMENTING AI

Chapter 10: TIPTOEING THROUGH THE MINEFIELD

Amazon ditched AI recruiting tool that favored men for technical jobs. (2018, October 10). *Reuters/The Guardian*. https://www.theguardian.com/technology/2018/oct/10/amazon-hiring-ai-gender-bias-recruiting-engine

Amazon scrapped 'sexist AI' tool. (2018, October 10). *BBC News*. https://www.bbc.com/news/technology-45809919

Baron, E. (2016, March 25). The rise and fall of Microsoft's 'Hitler-loving sex robot'. *Silicon Beat*. https://web.archive.org/web/20160325213945/http://www.siliconbeat.com/2016/03/25/the-rise-and-fall-of-microsofts-hitler-loving-sex-robot/

Gladwell, M. (2006). *Blink: The power of thinking without thinking*. Penguin Books.

Janis, I. L. (1972). *Victims of groupthink: A psychological study of foreign-policy decisions and fiascos*. Houghton Mifflin Company.

Kruger, J., & Dunning, D. (1999). Unskilled and unaware of it: How difficulties in recognizing one's own incompetence lead to inflated self-assessments. *Journal of Personality and Social Psychology, 77*(6), 1121–1134. https://doi.org/10.1037/0022-3514.77.6.1121

Larson, J. et al (2016, May 23). How we analyzed the COMPAS recidivism algorithm. *Pro Publica*. https://www.propublica.org/article/how-we-analyzed-the-compas-recidivism-algorithm

Thaler, R. H. & Sunstein, C. R. (2008). *Nudge: Improving decisions about health, wealth, and happiness*. Penguin Books.

Vartan, S. (2019, October 24). Racial bias found in a major health care risk algorithm. *Scientific American*. https://www

.scientificamerican.com/article/racial-bias-found-in-a-major-health-care-risk-algorithm/

Wiessner, D. (2023, August 11). Tutoring firm settles US agency's first bias lawsuit involving AI software. *Reuters.* https://www.reuters.com/legal/tutoring-firm-settles-us-agencys-first-bias-lawsuit-involving-ai-software-2023-08-10/

Chapter 11: COLLABORATION AND CHANGE MANAGEMENT AROUND AI

Case Study 1: The £10 billion IT disaster at the NHS. (2019, January 20). *Henrico Dolfing.* https://www.henricodolfing.com/2019/01/caschange-study-10-billion-it-disaster.html

Kotter, J. (2018). *8 steps to accelerate change in your organisation* (eBook). www.kotterinc.com/wp-content/uploads/2019/04/8-Steps-eBook-Kotter-2018.pdf

Luff, N. (2019, April 15, 2019). Is this Britain's most successful digital transformation? *Management Today.* https://www.managementtoday.co.uk/britains-successful-digital-transformation/innovation/article/1582047

The 8 Steps for leading change (n.d.). *Kotter Inc.* https://www.kotterinc.com/methodology/8-steps

The boat race: A perfect crew (2009, October 2, 2009). Cambridge Ideas [Video]. YouTube. https://www.youtube.com/watch?v=MXLg9nsuo9I

Why do most transformations fail? A conversation with Harry Robinson (2019, July 10, 2019). [Video]. *McKinsey.* https://www.mckinsey.com/capabilities/transformation/our-insights/why-do-most-transformations-fail-a-conversation-with-harry-robinson

Why so many digital transformation projects fail (2023, February 24, 2023). *Engineering.com.* https://www.engineering.com/story/why-so-many-digital-transformation-projects-fail

Chapter 12: AI IMPLEMENTATION

Dey, A. K. (2000, November). Providing architectural support for building context-aware applications [Scholarly project]. http://swiki-lifia.info.unlp.edu.ar/ContextAware/uploads/10/dey thesis.pdf

IDC (2024, August 19). *Worldwide spending on artificial intelligence forecast to reach $632 billion in 2028, according to a new IDC spending guide.* https://www.idc.com/getdoc.jsp?containerId=prUS52530724

Ransbotham, S. et al. (2020, October 20). 'Expanding AI's Impact with Organizational Learning.' MIT Sloan Management Review, Big Ideas Artificial Intelligence and Business Strategy Initiative [Website]. (Findings from the 2020 Artificial Intelligence Global Executive Study and Research Project.) https://sloanreview.mit.edu/projects/expanding-ais-impact-with-organizational-learning/

US Securities and Exchange Commission (2024, March 18). *SEC charges two investment advisers with making false and misleading statements about their use of artificial intelligence* [Press release]. https://www.sec.gov/newsroom/press-releases/2024-36

Chapter 13: INVESTING IN PEOPLE

De Smet, M. et al (2023, September 11). Some employees are destroying value. Others are building it. Do you know the difference? *McKinsey Quarterly.* https://www.mckinsey.com/capabilities/people-and-organizational-performance/our-insights/some-employees-are-destroying-value-others-are-building-it-do-you-know-the-difference?

Deloitte (2023). The Deloitte Global 2023 GenZ and Millennial Survey. https://www2.deloitte.com/cn/en/pages/about-deloitte/articles/genzmillennialsurvey-2023.html

Deloitte (2024). Deloitte's State of Generative AI in the Enterprise Quarter One report. https://www.deloitte.com/ce/en/services/consulting/research/state-of-generative-ai-in-enterprise.html

Dunlop, A. et al (2023, March 26). Hey bosses – Here's what GenZ actually wants at work. *Deloitte Digital.* https://www.deloitte-digital.com/us/en/insights/perspective/gen-z-research-report.html

Goleman, D. (1995). *Emotional intelligence: Why it can matter more than IQ.* Bantam.

Jordan, J. & Sorell, M. (2019, June 4). Why You Should Create a "Shadow Board" of Younger Employees. *Harvard Business Review.* https://hbr.org/2019/06/why-you-should-create-a-shadow-board-of-younger-employees

Majority of hiring managers favor candidates with AI skills over those with more experience (2014, January 15). *ResumeTemplates.* https://www.resumetemplates.com/majority-of-hiring-managers-favor-candidates-with-ai-skills-over-those-with-more-experience/

Mazzola, A. (Producer). *The Profit: An Inside Look with Marcus Lemonis* (TV series). CNBC.

Prioritise people: Unlock the value of a thriving workforce (2023, April 24). *Business in the Community.* https://www.bitc.org.

uk/report/prioritise-people-unlock-the-value-of-a-thriving-workforce/

Randstad (2024). Randstad Workmonitor 2024. https://www.randstadusa.com/landing/workmonitor-2024/

Senge, P. (1994). *The fifth discipline: The art and practice of the learning organization.* Random House.

CONCLUSION

Kotter, J. P., & Cohen, D. S. (2012). *The heart of change.* Harvard Business Review Press.

Maslow, A. H. (1943). A theory of human motivation. *Psychological Review, 50*(4), pp. 430–437. Washington, DC: American Psychological Association.

Pope, A. (1711). *An essay on criticism.* Project Gutenberg. https://www.gutenberg.org/files/7409/7409-h/7409-h.htm

Tobak, S. (2011, October 31). Facebook's Mark Zuckerberg – Insights for entrepreneurs. *CBS News.* https://www.cbsnews.com/news/facebooks-mark-zuckerberg-insights-for-entrepreneurs/

Winfrey, O. (2013). *Harvard 2013 Commencement Speech* [Video]. YouTube. https://www.youtube.com/watch?v=GMWFieBGR7c

Praise for Generation AI

Tony Evans' Generation AI masterfully bridges the gap between time-less wisdom and cutting-edge technology. This book is a game-changer for leaders navigating the AI revolution and GenZ confusion. Evans' insights on uniting generations and leveraging diverse perspectives are pure gold. As someone passionate about finding meaning at work, I can confidently say this book is a blueprint for creating purposeful, high-performing multi-generational teams in the AI era. It's not just a must-read; it's a must-implement for any leader serious about thriving in the future of work. Tony Evans proves once again why he's a thought leader at the cross-sections of tech and team in business.

Dr. Corrie Jonn Block,
Chief Executive Coach and Bestselling Author of Love@Work.

As a leader of a multi generational workforce, navigating the AI revolution needs people and collaboration. This book is an excellent guide for all modern, forward thinking leaders

Jon Ghazi,
Chief Growth Officer and
CEO EMEA, Annalect, Omnicom Group.

Tony Evans has done something miraculous with "Generation AI"–he's made AI accessible, practical, and deeply human. For those of us committed to building strong workplace cultures, this book is a revelation. It goes beyond the hype, providing leaders with the knowledge and confidence to navigate AI's impact on teams, trust, and collaboration. It's essential reading for anyone looking to future-proof their organisation while keeping people at the centre of progress.

Lucy d'Abo,
CEO & Founder of Together Incorporated, the first dedicated workplace culture consultancy in the Middle East

As the world is undergoing a major paradigm shift brought on by the advances in AI. Tony provides excellent guidance in how to navigate this process while bridging the generational and technological gaps. A must read for all current and aspiring leaders.

Nayab Rafiq. CEO Pinnacle Middle East

Tony Evans convincingly argues that a collaborative partnership between Gen Z and Boomers is essential for successfully implementing AI in organisations. We are shaped by the contexts in which we grew up–WHEN and WHERE. This book explains how leaders can effectively leverage these differences to benefit the organisation during a change process.

Michael Schachner,
MSc. Head of Research at The Culture Factor Group

Extras

www.ingramcontent.com/pod-product-compliance
Lightning Source LLC
Chambersburg PA
CBHW040915210326
41597CB00030B/5083